THEORIES AND TECHNOLOGIES

OF
SPRINKLER IRRIGATION
AND
MICRO-IRRIGATION EQUIPMENT

喷微灌装备理论与技术

李红 (Li Hong)
蒋跃 (Jiang Yue)
汤玲迪 (Tang Lingdi)
汤攀 (Tang Pan)
编著

江苏大学出版社
JIANGSU UNIVERSITY PRESS
镇 江

图书在版编目(CIP)数据

喷微灌装备理论与技术＝Theories and Technologies of Sprinkler Irrigation and Micro-irrigation Equipment：英文／李红等编著. —镇江：江苏大学出版社，2020.12
ISBN 978-7-5684-1417-3

Ⅰ. ①喷… Ⅱ. ①李… Ⅲ. ①喷灌—灌溉机械—教材—英文 Ⅳ. ①S275.5

中国版本图书馆 CIP 数据核字(2020)第 231723 号

喷微灌装备理论与技术
Theories and Technologies of Sprinkler Irrigation and Micro-irrigation Equipment

编　　著/李　红　蒋　跃　汤玲迪　汤　攀
责任编辑/郑晨晖
出版发行/江苏大学出版社
地　　址/江苏省镇江市梦溪园巷 30 号(邮编：212003)
电　　话/0511-84446464(传真)
网　　址/http://press.ujs.edu.cn
排　　版/镇江文苑制版印刷有限责任公司
印　　刷/镇江文苑制版印刷有限责任公司
开　　本/787 mm×1 092 mm　1/16
印　　张/10.75
字　　数/395 千字
版　　次/2020 年 12 月第 1 版
印　　次/2020 年 12 月第 1 次印刷
书　　号/ISBN 978-7-5684-1417-3
定　　价/45.00 元

如有印装质量问题请与本社营销部联系(电话：0511-84440882)

PREFACE

Sprinkler irrigation and micro-irrigation technologies are two of the main water-saving irrigation methods all over the world. Compared with traditional surface irrigation methods, the evaporation of water in sprinkler irrigation and micro-irrigation systems is greatly reduced by using irrigation pipes so that the application efficiency is highter. Moreover, the working pressure and the flow rate in sprinkler irrigation and micro-irrigation systems can be controlled to meet different requirements of irrigation quota and irrigation uniformity for different users. And the irrigation area and the irrigation duration can then easily be adjusted to achieve "precision irrigation". Meanwhile, they can coordinate perfectly with the modern agriculture and show a high degree of mechanized operation. Considering the operation convenience and cost of different irrigation methods, sprinkler irrigation and micro-irrigation technologies have the advantages of high-efficiency, water and fertilizer saving, production and income increasing, and management facilitating, hence they are important to the sustainable development of agriculture. With the purpose for the development of these technologies in poverty alleviation and agricultural modernization in China, and the purpose for the popularization and application of them in different countries, this textbook is mainly focused on studying the component design and selection of proper sprinkler irrigation and micro-irrigation systems. The main contents include: ① to introduce the international and domestic development situation and prospects of sprinkler irrigation and micro-irrigation technologies and equipment;

② to clarify the structures, operating principles, and design theories of the sprinkler irrigation and micro-irrigation equipment; ③ to evaluate these systems and their components so as to give recommendations for the application of different types of sprinkler irrigation systems in different projects; and ④ to cast some light on the selection of automation irrigation systems.

In Chapter 1 basic concepts and characteristics of sprinkler irrigation and micro-irrigation equipment are given. Meanwhile, the advantages and disadvantages of sprinkler irrigation and micro-irrigation respectively are discussed. Finally, the international and domestic development situations and prospects of sprinkler irrigation and micro-irrigation technologies and equipment are introduced.

In Chapter 2 typical sprinklers used in sprinkler irrigation and micro-irrigation are introduced. A sprinkler is one of the most important parts of a sprinkler irrigation system or a micro-irrigation system. According to the working pressure and the wetted radius, sprinklers can be categorized as micro-pressure sprinklers, low-pressure sprinklers, medium-pressure sprinklers and high-pressure sprinklers. According to the structure and the spray characteristics, they can be mainly categorized into four types: rotating (fluidic) sprinklers, fixed (refraction, diffusion) sprinklers, spraying porous tubes, and pulsating sprinklers. Meanwhile, the structures and design theories of these kinds of sprinklers are investigated. Finally, the test method for the characteristics of sprinklers are provided.

In Chapter 3 several types of small-scale sprinkler irrigation systems are introduced. A small-scale sprinkler irrigation system is usually featured by its small coverage in one irrigation cycle, simple control, and high mobility. It is mainly categorized as hand-held, hand-lift, trolley-type or tractor-driven system. After the introduction of the structures of these systems, the evaluation criteria and the hydraulic calculation method for the design of pipelines in a small-sized light sprinkler irrigation system are proposed, and the optimization model of this type of system is constructed.

In Chapter 4 the types of large and medium-sized sprinkler irrigation systems are introduced. These systems include the reel type sprinkler irrigation system, the side-roll wheel sprinkler irrigation system, the center pivot sprinkler irrigation system, and the lateral move sprinkler irrigation system. The structure, working principle, performance and characteristic of each sprinkler irrigation system are clarified. In addition, the field planning and layouts of each sprinkler irrigation system are designed according to the size and the shape of irrigtion area, as well as the type of irrigation and its capacity.

In Chapter 5 the micro-irrigation systems and equipment are introduced. Micro-irrigation methods can be classified as drip irrigation, micro-spray irrigation, small pipe irrigation, and infiltration irrigation according to different types of emitters and different forms of outflow. The quality of an emitter directly affects the service life and the irrigation quality of the micro-irrigation system. Therefore, the main technical parameters and performance indexes of common emitters are provided. Moreover, the pipelines and fittings, control, measurement and protection devices, filter equipment, and fertilizer apparatus are introduced.

In Chapter 6 the full-automation irrigation system and semi-automation irrigation system are introduced. In both irrigation systems, the typical components mainly include the central controller, the automatic valves, and the wires. Besides this, the line connection of the automatic control system is illustrated.

This book can be used as a reference for engineers and technicians engaged in the research of the sprinkler irrigation and micro-irrigation technologies and the water-saving irrigation projects, as well as the teachers and students who major in agricultural engineering in colleges and universities.

This book was written by Li Hong, Jiang Yue, Tang Lingdi and Tang Pan, and coedited by Tu Qin, Chen Chao, Liu Junping, Zhu Xingye and Xia Huameng. All the people contributed to this book are gratefully acknowledged. We acknowledge that this book was financially supported by the National Natural Science Foundation of China (No. 51679109, 51939005, 51809119), the National key Research and Development Project of China (2017YFD0201502), the Project Funded by the Priority Academic Program Development of Jiangsu Higher Education Institutions (No. PAPD-2018-87), and Agricultural Equipment Department Project of Jiangsu University.

Due to limited knowledge and time, the book might still have some limitations and deficiencies. We welcome any comments and suggestions from our readers.

The authors

CONTENTS

CHAPTER 1
Introduction

1.1 Basic concepts and characteristics of sprinkler irrigation and micro-irrigation equipment

1.1.1 Basic concepts of sprinkler irrigation

1. Sprinkler irrigation

Sprinkler irrigation is an advanced irrigation method which provides the necessary moisture conditions required for the normal growth of crops by using water pumps or conveying water raised to a height into irrigated fields through pressure developed in pipelines. Water is sprayed into the air through the nozzle of the sprinkler and spread uniformly across the field in the form of droplets, imitating rainfall.

2. Sprinkler irrigation equipment

Sprinkler irrigation equipment, or commonly referred to as sprinkler irrigation machinery (or machines and tools), is one of the main means of spraying the irrigation water. Depending on its function, equipment can be divided into the following components[1]:

① Spraying equipment (sprinkler and spray hole tube);

② Water conveyance equipment (pipelines and accessories for sprinkler irrigation);

③ Supercharging equipment (includes water pumps, power machine, and transmission equipment);

④ Walking and supporting equipment (includes power machine, transmission equipment, road wheel, girder, each kind of joints, and draught attachment, etc.);

⑤ Control equipment (running and brake devices, safety equipment, walking synchronization, pressure adjustment, sensor, automatic valves, operation box, etc.);

⑥ Measurement equipment (vacuum meter, pressure gauge, timer, soil moisture sensor, electric instrument, thermal meter, etc.);

⑦ Accessory equipment (fertilizer and pesticide injector, filtration unit, etc.).

3. Sprinkler irrigation machine

The sprinkler irrigation machine combines the power machine, water pump, pipelines, sprinkler, and shifter which can be assembled in a variety of ways.

There are many kinds of sprinkler irrigation machines. Based on the mode of operation, the sprinkler irrigation machines can be classified into two types: a traveling sprinkler irrigation system and a fixed sprinkler irrigation system. There are two different types due to the differences in the systems' covered areas and characteristics. The fixed sprinkler irrigation system includes a trolley sprinkler irrigation system, an end tow sprinkler irrigation system, and a side-roll wheel sprinkler irrigation system. The traveling sprinkler irrigation system includes a reel type sprinkler irrigation system, a center pivot sprinkler irrigation system and a lateral move sprinkler irrigation system.

4. Sprinkler irrigation system and its classification

A sprinkler irrigation system is a type of water conservancy facility which connects an irrigation water source and sprinkler equipment to a field to spray irrigation water uniformly and ensure the crops are treated properly.

The source of water for a sprinkler irrigation system could be a river, channel, lake, pond, wellspring, etc. The source of water should meet the specific quality and quantity requirements of sprinkler irrigation. Field engineering in sprinkler irrigation systems includes water conveyance pipes, the structure of pipes, and land formation. The land formation should be geometrical and concur with specific planning and design.

Because there are several types of sprinkler irrigation systems, their classifications are abundant. They are specified below.

(1) Classification by the system acquires working pressure

① Mechanical pressure sprinkler irrigation system. The system acquires pressure by a pumping machine.

② Gravity pressure sprinkler irrigation system. The system acquires working pressure by natural height.

(2) Classification by spray characteristics of the system

① Fixed sprinkler irrigation system. Sprinkler irrigation equipment, such as the nozzle, sprays water at one point. Examples are the pipeline sprinkler irrigation system and the fixed unit sprinkler irrigation system.

② Travelling sprinkler irrigation system. This system is made up of center pivot sprinkler irrigation system or a lateral move sprinkler irrigation system. The sprinklers spray the water during movement.

(3) Classification by the composition of system equipment

① Pipeline sprinkler irrigation system. In this system, the water source, sprinkling irrigation pump, and each sprinkler are connected by a single-stage or multistage pressure pipes. It is called a pipeline sprinkler irrigation system because the pipeline is the most important part of the system. Depending on the degree of movement of the pipe, the system is classified as a fixed pipeline sprinkler irrigation system, a half-fixed pipeline sprinkler irrigation system, and a movable pipeline sprinkler irrigation system.

② Unit sprinkler irrigation system. This system takes the sprinkler irrigation machine (unit) as the main part. It is divided into a fixed unit and traveling unit sprinkler irrigation systems.

1.1.2 Advantages and disadvantages of sprinkler irrigation and its application range

1. Advantages of sprinkler irrigation

(1) Water conservation

There are two different occasions when water may be lost during irrigation. The first one may occur during the process of water conveyance from the water source to the field. The second one is during the irrigation progress itself.

In China, for example, the effective utilization coefficient of irrigation water in farmland is 0.52, with more than half of it wasted in irrigation. Sprinkler irrigation uses a pipeline to convey and distribute water, and thus the water loss is minimal and uniformly distributed in the irrigated area, without noticeable surface runoff or erosion. In comparison to the traditional surface irrigation methods, the utilization coefficient in sprinkler irrigation could be more than 0.8, which saves about 40% of the irrigated water.

(2) Increased crop yield and improved crop quality

The effectiveness of sprinkler irrigation is advantageous in yielding higher crop production. Sprinkler irrigation can meet crop water requirements with the appropriate amount and frequency of water, control and prevent soil moisture deficits, and maintain soil fertility. It may also adapt to different water requirements depending on the growth stage of the crops. Sprinkler irrigation, resembling rainfall, makes soil moist, upkeeps the soil granular structure, and creates healthy soil conditions for the root growth of crops. Generally, the yield of field crops could be increased by 10% to 20%, whilst economic could be increased by about 30%, and vegetable crops could be increased by 1 to 2 times.

Water delivery in sprinkler irrigation commonly occurs underground. Thus, the occupation of ridges and channels is reduced, and the utilization rate of the farmland can be increased by 5% to 7% since there is more room. Sprinkler irrigation can change the microclimate in the irrigated area, which increases the humidity of the surrounding air, and adjusts the temperature difference between day and night. Sprinkler irrigation not only prevents the damage of the dry-hot wind and frost which may inflict onto crops, but can also improve the quality of the crops significantly.

(3) Labor-saving

Sprinkler irrigation has a high mechanization degree, which reduces the labor requirements of irrigation and improves working efficiency. Thus, it avoids building up the field ridge and the channels every year, leading to a decreased necessity of labor. Compared to the traditional method of surface irrigation, sprinkler irrigation can improves the efficiency of labor about 20 to 50 times.

(4) Adaptable

A prominent advantage of sprinkler irrigation is that it can be used in all types of soil and crops, and is mostly flexible to all topographic conditions. For example, sprinkler irrigation can be used in areas where the ground is undulating and in places that irrigation systems may be difficult to install, such as in areas where the slope has more than a 5% grade, or where there is sandy soil. In an area with a high water level, ground irrigation makes the soil layer wet, causing salinization in the soil; sprinkler irrigation can be used to regulate the water condition of the upper layer of soil, therefore avoiding the occurrence of salinization. Due to the low demand for labor, sprinkler irrigation can save a large amount of time and money on farmland consolidation.

2. Disadvantages of sprinkler irrigation

(1) The spraying operation is affected by wind drift

During sprinkling, wind drift tends to sway away any water droplets; such off-target movement and deposition are often called spray drift. Wind drift is apt to raise the amount of water lost and reduce spray efficiency, which also changes the shape of the water distribution pattern and sprays distance. This results in a reduction in the uniformity and quality of irrigation. Hence, it is unfavorable for sprinkler operations when the wind velocity is higher than level 3 km/s.

(2) High equipment investment

Due to the need for a variety of equipment and pipes for the sprinkler irrigation system, the pressure of the system is high, and the requirements for the pressure resistance of the equipment are high. As such, the investment of a sprinkler irrigation system is also high. For example, the investment of fixed pipeline sprinkler irrigation system is from $900 to $1 200 per acre, whereas the investment of a half-fixed pipeline sprinkler irrigation system is from $300 to $450 per acre, the investment of reel type sprinkler irrigation system is from $400 to $600 per acre, and the investment of large-sized sprinkler irrigation units is from $500 to $800 per acre.

(3) Energy dissipation

Pumping water through channels or pipelines into a field is the working process of surface irrigation, which can be conducted through automatic irrigation. However, automatic sprinkler irrigation uses water pressure to break the spray into droplets and distribute them within the prescribed limits. This requires more energy in comparison to traditional surface irrigation methods. To save energy in the wake of rising costs, sprinkler irrigation can be developed to have lower pressure. Combining sprinkler irrigation with drip irrigation technology can further develop micro-sprinkler irrigation and save energy.

3. Conditions for the application of sprinkler irrigation

As is mentioned above, sprinkling irrigation is an advanced irrigation method, which has a wide range of application and can be used for almost all crops under various topography and soil conditions. Additionally, it can be used for spray fertilizer while preventing frost, heat, dust, etc. But, it also has the disadvantage of not being able to use sprinkler irrigation in any given area. To take full advantage of sprinkler irrigation and minimize its disadvantages, the following situations need to be taken into consideration when deciding whether to adopt sprinkler irrigation to achieve better efficiency.

① Local sources of capital are adequate and the crops with high economic benefits are cultivated.

② The terrain is undulating or steep, making it difficult to install ground irrigation.

③ The soil permeability is strong, or the infiltration rate is high.

④ Irrigation crops requiring regulation of the micro-climate in the field, including heat or frost protection.

⑤ Not the windy area or the irrigation monsoon is minimal.

1.1.3 Basic concepts of micro-irrigation

1. Micro-irrigation

Micro-irrigation is a local irrigation technology. Micro-irrigation equipment conveys pressured water into fields, so the soil near the root of the crop is moistened by a water irrigating device.

2. Micro-irrigation equipment

The micro-irrigation equipment consists of pump, strainer, fertilizer apparatus, control equipment, measuring equipment, protective equipment, water pipes and fittings, emitter, etc. (See Chapter 5 for more details.)

3. Micro-irrigation system and classification

The typical micro-irrigation system consists of four parts: a water source project, head works, water distributing network, and emitter[2].

(1) Water sources project

Micro-irrigation water sources could draw from a river, channel, lake, reservoir, well, spring and so on.

(2) Head works

Head works include a water pump, power machine, filter plant, fertilizer apparatus, controller, control valve, inlet, exhaust valves, measuring instrument, etc. Its function is to draw water from the water source by supercharging and dealing it into water flow so that it meets the requirements of micro-irrigation and conveys it to the irrigation delivery and distribution pipe network.

(3) Water distributing network

This network includes manifolds, laterals, safety control and adjustment devices. Its function is to distribute and transfer water from the head works to the drip irrigation emitters.

(4) Emitter

The emitters irrigate water directly and are made up of dippers, drip irrigation hoses, minisprinklers, foaming sprinklers, small tube flow devices, etc. The purpose of emitter is to reduce water pressure inside pipes, turn the water flow into a trickle, or drips, and spray it into the soil.

Depending on the water outlet form of the emitter, micro-irrigation is categorized as drip irrigation, micro-sprinkler irrigation, and small tube flow irrigation.

(1) Drip irrigation

Drip irrigation is an irrigation technology that uses drip irrigation emitters, such as dippers and drip irrigation hoses, to wet the soil near the root area of the crop in the form of dripping water or small streams of water. The flow rate of an emitter is usually no more than 12 L/h; the most common flow rate of an emitter is from 1 to 4 L/h. When the emitter is placed on the ground, it is called surface drip irrigation. When the emitter is placed under the ground and conveys the water to the root area of the crop under the surface directly, it is called under subsurface drip irrigation. The irrigation method which puts drip irrigation hoses under agricultural film is called drip irrigation under mulch.

(2) Micro-sprinkler irrigation

Micro-sprinkler irrigation is one of the irrigation methods which uses drip irrigation emitters such as minisprinklers, micro-sprinkler irrigation hoses to spray the pressure water in the form of spray water on the soil surface near the root area of the crop; it is commonly referred to as microspray for short. The flow rate of the minisprinkler is usually no more than 250 L/h, and the common flow rate of the minisprinkler is from 20 to 240 L/h.

(3) Irrigation with small tubes

Irrigation with small tubes is another irrigation method that involves the usage of a regulator to steady water flow, small tubes to disperse water flow. This fills the soil surface with a small stream of water. The flow rate of the irrigation with small tubes is equivalent to the flow rate of a micro-dripper.

1.1.4 Advantages and disadvantages of micro-irrigation and its application range

1. Advantages of micro-irrigation

(1) Saving water

Because of the minute flow rate of micro-irrigation, it can precisely determine surface irrigation and meet the requirement of crop water requirement at any time, while avoiding the production of deep seepage. Most types of micro-irrigation are local, which only wet the soil surface near the roots. Thus, soil evaporation is small, leading to the conservation of more water.

(2) Uniform irrigation

A micro-irrigation system can control the water flow rate of each irrigation device effectively. The irrigation is therefore very evenly distributed, reaching between 80% to 90% uniformity.

(3) Increase in production

Micro-irrigation can apply water, fertilizer, and pesticide to plants with the appropriate amount and frequency. This improves the efficiency of the substances, reduces the occurrence of plant diseases and insect pests, inhibits the growth of weeds, keeps the soil aggregates together, provides good conditions for crop growth to increase production, and improves product quality. Since micro-irrigation is so advantageous, it raises crop production to have 15%~40% more output than any other irrigation method.

(4) Strong adaptability

Depending on the infiltration characteristics of the soil, suitable drip irrigation emitters can be selected to adjust the efficiency of micro-irrigation, so that surface runoff and leakages are not produced. Micro-irrigation can work effectively in any complex terrain, even for crops cultivated on very steep or on rocky patches of land.

2. Disadvantages of micro-irrigation

(1) The drip irrigation emitters can become blocked.

The small flow channel plugging of water-drippers is the main problem of drip irrigation. Further, it is difficult to detect the location of the blockage in the channel, obstructing the water in the capillary tube, and possibly jeopardizing the entire system if the problem becomes severe. The main causes of blockage include physical, biological and chemical factors, such as silt, organic matter, microorganisms, and chemical precipitate in the irrigation water. All of these factors require water to have the highest quality possible, as so many things can go wrong. Therefore, it is vital that water is filtered and that it undergoes precipitation and chemical treatment when necessary. Fortunately, at present, a miniature type of dripper with a large port, small flow rate, and good anti-jam performance has been developed. The problem is gradually being solved as research improving the efficiency of drip irrigation is conducted.

(2) Salt accumulation

When drip irrigation is utilized on soil with high salinity or saline water, the salt accumulates around the edges of the moist soil, and after rainfalls, the salts may penetrate the root zone of the crop, leading to soil damage. It should be leached with additional spray or one must use ground irrigation. It is advisable not to use drip irrigation on soils with high salinity or with saline water, and where there is no adequate flushing condition or adequate rainfall.

(3) High initial cost

Compared with ground irrigation systems, drip irrigation systems require more equipment and materials, so it costs more.

3. Application range of micro-irrigation

Micro-irrigation has very important applications. Mainly, it is used to minimize water loss through evaporation and raise the utilization rate of water, which means producing higher quality crops with the least amount of water possible.

Because the investment of micro-irrigation is much higher, most countries do not apply micro-irrigation systems to their field crops, but rather to more economic crops such as fruit trees, vegetables, flowers, and medicinal herbs. Therefore, farmers can get better economic and social benefits with their investment as soon as possible.

1.2 International and domestic development situations and prospects of sprinkler irrigation and micro-irrigation technology and equipment

1.2.1 Development situations and prospects of sprinkler irrigation

1. Present situation and trends in foreign sprinkler irrigation

Foreign applications of sprinkler irrigation technology and projects revolutionized before the application in China. Therefore, the systems are more developed abroad. Sprinkler irrigation was first emerged in the 19th century, when Germany, Italy, the

United States, and the former Soviet Union, were users of the first generation of sprinklers in (self-pressured sprinkler irrigation) in 1917, but until 1920 the application was limited to irrigating vegetables, nursery gardens, and orchards. The earliest use of spray irrigation in foreign countries was the stationary system. In the 1930s, there were rotating sprinklers and sprinkler irrigation machines. Following 1930, they were gradually improved in the southern United States, England, France, Germany, and Italy. After World War Ⅱ, end tow sprinkler irrigation systems (1948), side-roll wheel sprinkler irrigation systems (1951), center pivot sprinkler irrigation systems (1955), reel type sprinkler irrigation systems (1966), lateral move sprinkler irrigation systems (in the late 1960s), and other large and medium-sized sprinkler irrigation systems emerged in succession, so that the spray irrigation technology developed rapidly all over the world [3].

At present, foreign sprinkler irrigation technology shows the following trends[4-7]:

(1) Diversified development of sprinkler irrigation equipment and sprinkler irrigation system. Different countries and regions apply different sprinkler irrigation equipment and systems. Therefore, each country develops a variety of sprinkler equipment and sprinkler systems according to their dynamics. The United States vigorously promotes the center pivot sprinkler irrigation system and lateral move sprinkler irrigation system; these two kinds of systems suit the current situation of farms in the United States and are now widely used in many countries. In recent years, Russia developed the boom sprinkler irrigation system, introducing a large center pivot sprinkler irrigation system called a "Valley" from the United States. Russia utilized that design when developing a center pivot sprinkler irrigation system called a "Ferrigator", and a side-roll wheel sprinkler irrigation system called a "Walerica". Germany, France, Australia, and other countries focus on developing the reel type sprinkler irrigation system.

(2) Enlarging the control area of a single machine and overall system, and improving the adaptability of the unit. Foreign countries expand a single-machine control area to improve the production efficiency of the sprinkler irrigation machine, reduce the equipment investment, and cut labor and operation management expenses. In the United States, Nebraska, the Lekard company has developed the world's largest center pivot sprinkler irrigation system. In order to improve the production efficiency of land utilization and adapt to large sprinkler irrigation machine marshalling operation (generally from 2 to 10), the current irrigation system in the United States has been developed from being a single block between 240 to 960 acres in the 1950s to a large area of irrigated area of 12 000 to 60 000 acres. While expanding the control area, they attempted to improve the adaptability of the unit; the climbing ability of the irrigation system was increased from 6% to 30%. The truss structure has also been developed from the steel structure in the past to the present aluminum alloy structure. Because of this, the frame weight was reduced by 60%.

(3) Saving energy. Due to the rising energy costs all over the world, some countries have to inhibit the development speed of sprinkler irrigation. According to the Nebraska State Investigations, irrigation accounts for 43% of the energy consumed in agriculture, while sprinkler irrigation consumes 10 times more energy than other farming measures. To improve the development potential of sprinkler irrigation, it is necessary to take the

path of energy conservation. There are two main energy-saving measures adopted at present:

① Development of the low pressure sprinkler. In the United States, the high-pressure sprinkler was replaced with a low-pressure sprinkler of 245 kPa, saving $9.6 per 1 000 m^3 of water. The former Soviet Union developed a double cantilever irrigator with a working pressure of 100 kPa. In the United States, during the state of California's international drainage equipment exhibition in 1982, there were more than 100 kinds of low-pressure sprinklers developed by each country. In particular, the newly developed non-circular nozzles (square and rectangular nozzles) had a good water distribution in a low pressure, which became utilized by many people. After the 1990s, low-pressure sprinklers have been used in large traveling sprinkler irrigation machines produced by each country.

② Harnessing new energy, such as wind energy, biogas energy, solar energy, etc. In terms of sprinkler irrigation, wind power is the most promising energy. Wind energy has already been used in the United States, Australia, and other countries. The large plain region in America has 120 million acres of irrigated area, and more than half of it used wind pumps in the late 20th century.

(4) Extensive use of light-weight pipes and plastic pipes. To reduce the materials used, the weight of the unit, and labor intensity, while improving the work efficiency and reducing the investment costs of the irrigation pipelines, light-weight pipes, and plastic pipes are widely used all over the world. In many countries, more than two-thirds of the pipes used are usually plastic pipes. Due to the large use of cheap plastic pipes, irrigation systems in many developed countries have already disregarded channels and used pipelines to deliver irrigation water. For example, in countries such as Israel, France, and Spain, the water pipeline network is spread across the entire country's land. Farmers can only purchase water-saving irrigation equipment to connect with the main pipe network system so that water is diverted into the fields.

(5) Adoption of automation techniques. With the development of the computer, automatic technology such as microcomputer control is adopted in sprinkler irrigation systems, saving labor, improving spraying quality and production efficiency, and guaranteeing the operation reliability of the unit.

(6) Development of comprehensive utilization with the specific national economic departments. The comprehensive utilization of sprinkler irrigation equipment, work on multi-project operation and making full use of the multifunctional economic benefit of sprinkler irrigation equipment. For example fertilizer, spray pesticides and herbicides; protection against the frost and hot air; industrial and sports ground dust removal, concrete construction maintenance, factory cooling protection from heat; sprinkler irrigation on horticulture, flower, and lawn; city fountain, etc.

(7) Paying attention to the basic theory of sprinkler irrigation and the development of new technologies. Paying attention to the development of the basic theory, new technology research of sprinkler irrigation, the multidisciplinary cooperation and theory research of sprinkler irrigation lays the foundation for the perfection and improvement of hydraulic performance and mechanical properties of sprinkler irrigation equipment. It also provides a reliable criterion and basis for the planning and design of the sprinkler irrigation system reasonably. Many countries, like the United States, Russia, and Japan

have been assigning great importance to basic theoretical research for years and have achieved some results, such as jet fission principle raindrop impact energy, hydraulic characteristics of the pipe network, soil infiltration theory, and so on.

2. Development status and trends in domestic sprinkler irrigation

The study of irrigation technology in China began in the 1950s and 1960s, but it revolutionized in the 1970s. The development of sprinkler irrigation in China has experienced the following four stages[8,9]:

First, scientific research and experimental trial. This stage started in 1954, where Shanghai, China adopted the refraction sprayer for vegetable crops for the first time in the 1960s. China developed the worm gear-type sprinkler successfully and it was suitable for irrigating vegetable crops in Wuhan and other places. Hubei, Hunan, Guangdong, Jiangxi, Fujian, Sichuan and Liaoning provinces, as well as other provinces, had also successively tested the application for the economic benefits and the field crops.

Second, the climax of technological and equipment development. This stage lasted about 10 years, ranging from the mid-1970s to the mid-1980s, and was the stage of the rapid development of sprinkler irrigation technology in China. Among them, in 1976, the Chinese Academy of Sciences and Hydro Power Ministry listed sprinkler irrigation as a key research project in 1978. The State Planning Commission listed sprinkler irrigation as its main promotion project, and the State Council listed sprinkler irrigation as one of the 60 key promotional programs in the country. During the period, almost all kinds of irrigation machines in the world were introduced. Also, the aluminum welded pipe production line, thin-walled steel tube production line, the ZY rocker sprinkler production line, and many more were introduced.

Third, wandering and low tide. This stage began in the mid-1980s, the main reason is that fundamental changes had taken place in China's agricultural system: the full implementation of the household contract responsibility system. The original sprinkler irrigation engineering could not adapt to this kind of system, and thus the development of sprinkler irrigation had been greatly impacted. Also, some areas used bamboo and stone tube as the sprinkler pressure pipeline, palter with low-level manufacturing of irrigation machines and equipment, which caused the failure of existing projects and sprinkler irrigation machine, meaning it couldnot operate. It resulted in a significant negative impact on the development of sprinkler irrigation.

Fourth, the recovery and steady development stage. Since the mid-1990s, the global demand for water is rising, and China is facing a serious challenge of water shortages. The northern water table of China is severely reduced and the water resources environment is deteriorating. To this end, China developed water-saving irrigation as basic state policy, and sprinkler irrigation as part of water-saving agriculture is included in the national development plan, thus entering the stage of steady development[10-15].

China has a vast territory, a complicated terrain, a changeable climate and a large variety of soil and crops, all of which provide favorable conditions for the development of sprinkler irrigation. According to the statistics, the areas which are suitable for the development of sprinkler irrigation in China are 600 million acres. Over 66% of the cultivated land in our country is hilly slopes, and the cultivated land area is about 400 million acres; irrigation of these areas cannot be solved by traditional ground irrigation. After the use of sprinkler irrigation, the yield of China's economic crops increased and the interest is high, hence sprinkler irrigation is the main object of development.

1.2.2 Development situation and prospect of micro-irrigation[16]

1. Present situation and trends in foreign micro-irrigation

Drip irrigation evolved from underground irrigation. In 1860, the Germans adopted the clay pipe as the water seepage pipe for underground irrigation experimentation. The tube spacing was 5 m, buried depth about 0.8 m, and the tube was an outsourced 0.3~0.5 m thick filter layer. The experiment showed that crop yields had multiplied, which lasted more than 20 years. The public's interest was high. In 1920, a porous pipe was invented where water could flow through the holes in the wall of the water pipe and into the soil. After 1935, many experiments had been conducted on porous tubes made of different materials to test whether the quantity of the flow into the soil could only be regulated by the soil's moisture, rather than the water pressure in the piping system. Similar experiments were done in the Soviet Union (in 1923) and France, with no further progress.

With the development of industrialization, after the advent of plastic tubes, the formation of the drip irrigation system was promoted. At the start of the 1940s, the UK was the first to design a simple drip irrigation system with perforated plastic tubes, which was only used to irrigate flowers. Because the sizes of the perforated tube holes were not uniform, and the hole size changed over time, causing a large flow deviation, the simple holes in the wall of the water pipe were replaced by the drop head installed on the pipe. The first drops were very simple, formed by a hair tube wound around the tubular billet. In the late 1950s, Israel developed a long flow, injection-molding drop head. In 1960, they began to use drip irrigation systems for irrigating field fruit trees and greenhouse irrigation and achieved remarkable economic benefits. Since then, drip irrigation has become a new type of irrigation technology. In the 1970s, many countries gave a lot of attention to drip irrigation. It was introduced into commercial applications, becoming a popular new irrigation method in agricultural production.

To overcome the shortcomings of drip irrigation, Australia and the Soviet Union developed the micro-spray irrigation successfully. Then, while the Americans developed surge flow irrigation, China's irrigation with small tubes was also being designed. These irrigation methods are far more effective than the original category of drip irrigation. They formed new concepts of localized irrigation, but the basic characteristics of each of the irrigation methods are very similar: low operating pressure, small water flow rate, frequent irrigation, precise control of water flow rate, uniform irrigation, and only wetting part of the soil's root zone. Therefore, drip irrigation, surge flow irrigation, irrigation with small tubes, and small flow irrigation are called micro-irrigation.

Following the 1980s, researchers began to explore more water-saving underground drip irrigation techniques or SDI technologies for short. Currently, SDI technology has been applied to irrigation for maize, sugarcane, vegetable, fruit trees, and shrub irrigation in urban greening in Queensland in Australia, and California and Hawaii in the United States.

The International Commission on Irrigationand Drainage (ICID) micro-irrigation working group conducted four micro-irrigation surveys all over the world in 1981, 1986, 1991, and 2000. The latest survey showed that the world's micro-irrigation area had increased by 759% over 19 years, from 1981 to 2000. The micro-irrigation area of the United States had reached 1 050 000 hm^2, accounting for 28% of the total micro-irrigated area in the world and 4.91% of the total irrigated area in the United States.

Among the 88 member countries of the International Commission on Irrigation & Drainage (ICID), the combined micro-irrigation area is 3 749 154 hm^2, accounting for 1.5% of the total irrigated area of the member countries. Countries with more than 5% of the micro-irrigation area included Israel, Jordan, Cyprus, South Africa, Spain, Australia, France, as well as many others. As time went on, people gradually understood that micro-irrigation has been a kind of agricultural irrigation technology made for environmental protection and is an important part of modern agriculture, rather than simply another kind of agricultural irrigation technology. After the year 2000, people have placed much focus on promoting the development of the third world's agriculture, while developing and perfecting the micro-irrigation technology that has been formed. The emphasis was shifted to reducing investment in micro-irrigation systems and expand the using range of micro-irrigation.

2. Present situation and trends in domestic micro-irrigation

China pushed the development and application of drip irrigation technology from the introduction of the Mexican drip irrigation equipment in 1974. In the late 1980s to the early 1990s, China developed inter-tube emitters, micro-tubules emitters, orifice type emitters, distribute water emitters, refraction micro-jets, sand filters, screen filters, swirl sand separators, suction and exhaust valves, plastic pipes and pipe fitting, etc. Also, China proposed the concept of uniform drip irrigation and discovered it by simple energy dissipation micro-pipes creatively, improving the common formulas developed for calculating the head-loss of plastic pipes. China presented the calculation formula for the limited length of capillary pipes, developed some micro-irrigation techniques and equipment standards, and achieved a lot of valuable results about the water consumption of the micro-irrigation crops, irrigation regulation, and micro-irrigation design parameters. After the 1990s, China paid more attention to water-conserving irrigation. They introduced the foreign advanced production technology of drip irrigation pipes, drip irrigation belts, and pulse micro-irrigation equipment successively. China developed internally mounted drip irrigation hoses, pressure compensating emitters, rotating and refracting micro-emitters based on the understanding of imported technology. Also, China greatly progressed the development of quick coupling, anti-aging pipes, filter plants, fertilizer distribution equipment, and water conditioning progress, which narrowed the gap with foreign countries. At present, the country has already several companies distributed in the center of China, north China, Xinjiang and other places that produce five kinds of products, such as irrigation emitters, water-carriage pipes, purification filters, fertilizer devices, and control device, meeting the requirements of all kinds of users. China's micro-irrigation products developed rapidly in recent years; the variety and specifications are constantly increasing, and the performance of partial introduction technologies or production lines has reached the international level in the mid 1990s. However, most of the quality of self-developed micro-irrigation products still have a big gap in quality between similar foreign products. The performance is low and the product quality cannot meet the requirements completely[17].

Starting from The Ninth Five-Year Plan, China has carried out the construction of 300 water-saving, increasingly key counties and arranged a large number of water-saving and synergistic demonstration projects. The investment in water-saving irrigation gradually increased, which effectively promoted the rapid development of micro-irrigation. After 2000, China's micro-irrigation entered a new era of rapid development, especially in the development of drip irrigation cotton film in Xinjiang, which has greatly contributed to the increase of the national micro-irrigation area. By the end of 2009, the national micro-irrigation area had been counted as 1 666 000 hm^2.

Micro-irrigation is a modern agricultural technology which combines the mechanized irrigation with automatic irrigation organically. It is one of the most important technologies as it promotes regional agricultural economic development, increases farmers' income and speeds up the pace of agricultural modernization. It is an irreplaceable important part of modern agriculture. Also, micro-irrigation is adept in combating desertification, increasing soil and water conservation, and improving the ecological environment. With the progress of micro-irrigation technology, its application field will become more and more widespread and it will play an increasingly important role in future modern agricultural production[18].

References

[1] Li S Y. Theory and design of spray sprinklers[M]. Beijing: Weaponry Industry Press, 1995.

[2] Yuan S Q, Li H, Wang X K, et al. Irrigation technology and equipment of spray micro-irrigation[M]. Beijing: China Rural Water and Hydropower, 2014.

[3] Chen C. Brief history and development trend of the research on sprinkler system at home and abroad[J]. Drainage and irrigation machinery, 1999, 17 (2): 45-47.

[4] Guo H B, Shi Q. Brief introduction of water saving irrigation development at home and abroad[J]. Water Saving Irrigation, 1998(5): 23-25.

[5] Xu Z H, Sun J G. Discussion on energy saving of sprinkler irrigation project [J]. China Rural Water and Hydropower, 2000, 2: 28-29.

[6] Wu J S, Li J S, Li Y N. The key research field of high and new technology in water-saving agriculture in twenty-first Century[J]. Journal of Agricultural Engineering, 2000, 16(1): 9-13.

[7] Yin C X, Xu B H. A review of the development of sprinkler irrigation in China for fifty years. China Rural Water and Hydropower, 2003, 2: 9-11.

[8] Su D F, Li S Y. Current situation and prospect of water saving irrigation equipment in China. Drainage and Irrigation Machinery, 1997, 15(3): 22-26.

[9] Kang S H, Li Y J. Development trend and Countermeasures of water saving agriculture in China in twenty-first Century[J]. Journal of Agricultural

Engineering, 1997,3(4): 1-7.

[10] Li Y N. Several problems that should be paid attention to present stage of developing water saving irrigation in China[M]. Beijing: China Rural Water and Hydropower Press,2001.

[11] Pan Z Y, Liu J R, Shi W D. The present situation of light and small mobile sprinkler unit and its gap with foreign countries[J]. Drainage and Irrigation Machinery, 2003,21(1): 25-28.

[12] Lu G R, Li Y N. After joining the WTO, the prospect and countermeasures of the Sprinkler Irrigation equipment development in China[J]. Water Saving Irrigation, 2002,4: 29-31.

[13] Hou S J, Liang Z L, Li S Y. Development ideas of the manufacturing industry of sprinkler irrigation equipment in China [J]. Drainage and Irrigation Machinery, 2002,20(6): 26-28.

[14] Li Y N. Discussion on some problems of the development of sprinkler irrigation technology in China[J]. Water Saving Irrigation, 2002: 1-3.

[15] Lan C Y, Yi X T, Xue G N. Research and development status and development direction of sprinkler irrigation equipment in China [J]. Drainage and Irrigation Machinery, 2005,23(1): 1-6.

[16] Yao B. Micro-irrigation engineering technology[M]. Zhengzhou: The Yellow River Water Conservancy Press,2011.

[17] Tang L, Sun X P, Zhang W B. Progress in technology and equipment of micro-irrigation[J]. Journal of Ningxia Agricultural College, 2003,24(2): 82-85.

[18] Suggestions on promoting agricultural mechanization and the good and rapid development of agricultural machinery industry of the State Council[EB/OL].[2010-07-05]. http://www.gov.cn/zwgk/2010-07/09/content_1649568.htm.

CHAPTER 2

Sprinklers Used in Sprinkler Irrigation

2.1 Introduction

The sprinkler is one of the most important parts of the sprinkler irrigation system. The pressured water passes through the air, dispersing into tiny droplets and spreading evenly over the irrigated area; it is also known as the sprayer[1]. Its main function is to transform the pressured energy of the water current into kinetic energy to form droplets similar to raindrops and distribute it evenly to the irrigated area. The sprinkler can be installed on a fixed or moving tube, onto the delivery pipes of a composite truss of traveling sprinkler irrigation units, and on the traction frame of a hose reel sprinkler irrigation system. It consists of a complete sprinkler irrigation mechine or sprinkler irrigation system with matching machines and pumps. The performance and appropriate usage of a sprinkler play a decisive role in the spraying quality, amount of economic benefits, and functional reliability of a sprinkler irrigation system or a sprinkler irrigation machine.

2.1.1 Classification of the sprinklers

Sprinklers can be classified depending on their different methods of functionality. For example, according to the working pressure (or wetted radius) of the sprinkler, the classification of the work characteristics and materials generally includes the following two categories[2].

1. Working pressure and wetted radius

According to the working pressure and wetted radius, generally the sprinkler will be categorized as a micro-pressure sprinkler, low-pressure sprinkler (or short wetted radius sprinkler), medium-pressure sprinkler (or medium wetted radius sprinkler) or high-pressure sprinkler (or long wetted radius sprinkler). The flow rate, wetted radius, characteristics, and application of the sprinkler are as shown in Table 2-1 below.

Table 2-1 The sprinklers classified according to working pressure and wetted radius

Classification	Working pressure/kPa	Wetted radius/m	Flow rate/(m^3/h)	Characteristics and range of application
Micro-pressure sprinkler	50～100	1～2	0.008～0.3	The minimum energy consumption, atomized, can be used in the micro-irrigation system and for irrigating greenhouse crops, flower, and gardens
Low-pressure sprinkler (or short wetted radius sprinkler)	100～200	2～15.5	0.3～2.5	Less energy consumption, the small water application rate of sprinkler irrigation, mainly used for the continuous self-walking spray sprinkler in a vegetable field, orchard, nursery, greenhouse, or park grassland
Medium-pressure sprinkler (or medium wetted radius sprinkler)	200～500	15.5～42	2.5～32	Good evenness, the moderate water application rate of sprinkler irrigation, a wide range of applications, such as a garden, grassland, orchard, vegetable field, meadow field crop, economy crop, and all kinds of soils
High-pressure sprinkler (or long wetted radius sprinkler)	>500	>42	>32	A wide sprinkler irrigation range, high productivity, high energy consumption, suitable for sprinkler irrigation of a big field and grass, both of which have a low requirement for sprinkling quality

2. Classification by structure and spray characteristics

According to the structure and spray characteristics of a sprinkler, it can be categorized into four types: a rotating (fluidic) sprinkler, fixed (refraction, diffusion) sprinkler, spray porous tube, and pulsating sprinkler.

(1) Rotating sprinkler

A rotating sprinkler circles around its plumb line at work. This sprinkler sprays using side rotation; water spouted from the nozzle shows a concentrated jet flow and is gradually broken under the action of air. Therefore, it has a wide wetted radius, large flow rate range, low water application rate, and high evenness. It is the basic form of a medium wetted radius and long wetted radius sprinklers, which is the most widely used type of sprinkler all over the world.

The driving mechanisms and reverse mechanisms are the most important parts of a rotating sprinkler. Therefore, based on the characteristics of the driving mechanism, the rotating sprinkler can be categorized as impact (impinging) sprinkler, impeller (worm gear) sprinkler, and reaction sprinkler. Among them, the impact sprinkler can be categorized into a fixed guide plate impact sprinkler, and wedge-guide impact sprinkler, depending on the form of the guide plate. The reaction sprinkler can also be categorized as a dial sprinkler, vertical swing arm sprinkler, and a total convection sprinkler[3-5] (fluidic sprinkler). Depending on whether there is a reverse mechanism and a specific number of nozzles, the rotating sprinkler is divided into the categories of a full circle spray sprinkler, fan spray sprinkler, single nozzle sprinkler, and double nozzle sprinkler.

(2) Fixed sprinkler

When spraying, the parts of the sprinkler have no relative motion. The water flow in the full circle or part of the circumference (sector) is scattered simultaneously, so the wetted radius is relatively close. The water application rate near the sprinkler is much higher than the average water application rate; the general atomization degree is higher, and it advantageously has a simple structure, therefore working reliably. Based on the characteristics of the structure and the spray, the fixed sprinkler can be categorized into three kinds: refraction sprinkler, gap sprinkler, and diffusion sprinkler.

(3) Spray porous tube

A spray porous tube is also called a pore pattern sprinkler, which is characterized by the flow of water in the tube along many isometric holes with small linguloid jets. The tube can be used to rotate the oscillating mechanism around the tube 90° with its water pressure. A spray porous tube usually consists of one or a couple of small diameter tubes. In the upper part of the tube, a column or multi-column jet hole is arranged; its aperture is only 1 to 2 mm. Depending on the distribution of the water jet orifice, it can be divided into a single column and multi-column sprays.

Spray porous tubes have a simple structure, low working pressure, and convenient operation. However, its water application rate is high because the jet flow is small, so the influence of the wind is greater, the adaptability of the terrain is poor, the tube holes are easily blocked, the inner water pressure is strongly influenced by the terrain's ups and downs, and generally only applicable to flat land. Therefore, the application of large scale promotion has been relatively few, and it is only used in fixed fields such as greenhouses.

Currently, the most commonly used sprinklers in China are impact sprinklers, vertical impact sprinklers, fluidic sprinklers, and refraction sprinklers. Impact sprinklers and fixed sprinklers are the most relied upon.

2.1.2 Main structural parameters of the sprinklers

The main structural parameters of a sprinkler include four items: the inlet diameter of the sprinkler, nozzle diameter, sprinkler elevation angle, and connection size of a sprinkler and vertical pipe.

1. The inlet diameter of a sprinkler

The inlet diameter is the internal diameter of a sprinkler's hollow shaft or inlet pipe, and the unit is usually measured in mm. The inlet diameter is calculated by determining the reduction of hydraulic loss and structure compactness. After the inlet diameter of a sprinkler is determined, the water capacity and structure size can also be calculated. At present, the PY series sprinklers in China are named after the inlet nominal diameter. For rotating impact sprinklers, GB/T 22999—2008 "rotating sprinkler" stipulates that there are eight types of water inlet nominal diameter: 10, 15, 20, 30, 40, 50, 60, and 80 mm.

2. Nozzle diameter

The nozzle diameter is the smallest sectional diameter of a nozzle outlet, indicating the area of a nozzle runner and other sections; the unit is usually measured in mm. The nozzle diameter reflects the ability of a sprinkler to pass a flow under certain working pressure. In the scenario where the pressure is even, the larger the nozzle diameter, the larger the flow rate and the longer wetted radius, but the atomization strength of the sprinkler is relatively lower, and vice versa. For the nozzle diameter of a non-circular nozzle, it can be represented by the equivalent nozzle diameters, which can be calculated utilizing its flow rate.

3. Sprinkler elevation angle

The sprinkler elevation angle is the angle between the jet axis and the horizontal plane of the nozzle outlet. The unit is measured in degrees (°). In the case of the same working pressure and flow rate, the sprinkler elevation angle is the main parameter that affects the wetted radius and the water distribution. Selecting the right elevation angle results in the maximum wetted radius, which can reduce the water application rate and increase the spacing of the sprinkler pipe. This will facilitate the full use of a sprinkler, expand its coverage, reduce the pipeline investment in a pipe irrigation system, and reduce the total running cost of a sprinkler. The elevation angle is generally between 20° and 30°. The elevation angle of a large and medium-sized sprinkler is greater than 20°. The elevation angle of a small sprinkler is less than 20°, and the elevation angle of the commonly used sprinkler in China is 27° to 30°. To improve the resistance to wind, some sprinklers have used an elevation angle between 21° and 25°. Elevation angles of less than 20° are called low trajectory angles, which are generally used for under tree sprinkler irrigation and anti-frost sprinkler irrigation.

4. Connection size of a sprinkler and vertical pipe

The joint size between a sprinkler and vertical pipe vary. There are three kinds of connections: thread connections, quick joint connections, and flange connections. The PY series impact sprinklers in China adopt the pipe thread connection. The specification is given as: $\frac{1}{2}''$, $\frac{3}{4}''$, $1''$, $1\frac{1}{2}''$, $2''$, $2\frac{1}{2}''$, $3''$, $4''$.

2.1.3 Hydraulic performance parameters of the sprinklers

1. Pressure

The sprinkler pressure encompasses its working and nozzle pressure. Working pressure refers to the static water pressure measured at 20 cm from the sprinkler inlet when a sprinkler is working. The unit is measured in kPa, and the pressure gauge is generally installed on the vertical pipe. The nozzle pressure is the total pressure at the outlet of a nozzle (i.e., the velocity head). It can be used to evaluate the performance of the sprinkler. The working pressure of the sprinkler is nearly the same as the nozzle pressure. The pressure difference is the pressure loss of a sprinkler runner. The smaller the loss, the better the flow inside a sprinkler, and the higher the quality of the irrigation.

2. Flow rate

The flow rate of a sprinkler refers to the volume of water ejected from a sprinkler periodically; the unit is measured in m^3/h or L/min. The main factors that influence the flow rate of the sprinkler include the working pressure and nozzle diameter. In the case that the nozzle diameter matches or is close to the working pressure, the greater the working pressure, the greater the flow rate of the sprinkler, and vice versa.

The flow rate of a sprinkler can be measured using the volumetric method, weight method, or by the use of weirs and flowmeters, which can also be calculated by the hydraulic formula of the flow of the pipe, as follows:

$$Q=3\ 600\ \mu A\sqrt{2gH} \tag{2-1}$$

where Q is the flow rate of the sprinkler in m^3/h and μ is the flow rate coefficient of the sprinkler. It could be between 0.85 to 0.95. Generally, a big conical angle of a nozzle (such as 45°, 55°) is lower in value, and a small conical angle of a nozzle (such as 15°, 25°) is higher in value. A is the discharge area of the nozzle in m^2; g is the acceleration of gravity, 9.81 m/s^2; H is the working pressure head of the sprinkler in m.

3. Wetted radius

The wetted radius is the distance between the centerline of a sprinkler and the point in which a water application rate is a certain number under its normal rotating situation. For a sprinkler with the flow rate faster than 0.075 m^3/h, the water application rate of this point is 0.25 mm/h. For a sprinkler with a flow rate smaller than 0.075 m^3/h, the water application rate of this point is 0.13 mm/h. The reversing sprinkler can be measured at any angle, except at the maximum limit angle.

When a specific sprinkler's structural parameters have been determined, its wetted radius is mainly affected by the working pressure, wind speed, and rotational speed.

When the working pressure of a sprinkler increase, the wetted radius increases accordingly. This is possible when the working pressure is between a certain range that is specific to each sprinkler. Beyond that specific range of pressure, any increase in pressure will only increase the degree of atomization. The sprinkler wetted radius decreases when there is an increase in wind speed and rotational speed. Otherwise, the wetted radius will increase. Under the same condition of the flow rate, the larger the wetted radius, the smaller the water application rate of the sprinkler. The sprinkler spacing will also grow, which is beneficial for cost reduction and in improving adaptability. Therefore, the wetted radius is an important hydraulic performance indicator of the sprinkler.

The actual sprinkler wetted radius is calculated after the test. The test and calculation method of the rotating sprinkler wetted radius can be found in ISO 15886-3. There are many empirical formulas for estimating sprinkler wetted radius, and these are applicable at home and abroad.

(1) Kawaza formula

$$R=1.35(dH_0)^{\frac{1}{2}} \tag{2-2}$$

This formula applies to trajectory angles ranging from 30° to 32°.

(2) Lebedev's formula

$$R=\frac{H_0}{0.5+0.25\dfrac{H_0}{d}} \tag{2-3}$$

This formula applies to a trajectory angle of 30°, $800<\frac{H}{d}<4\ 000$.

(3) Nazarov formula

$$R=1.63H^{0.73}d^{0.17} \tag{2-4}$$

where R is the sprinkler wetted radius in m, d is the nozzle diameter in mm, H is the working pressure of the sprinkler in m.

It must be stated that the results of sprinkler wetted radills calculated by the above formulations are only in the absence of wind and without sprinkler rotation. In the case of wind and sprinkler rotation, the actual wetted radius is less than the calculated values of the above formulas. According to the statistics of several kinds of rotating sprinkler test data in China, the following formulas are recommended:

$$R_{PY_1}=1.70d^{0.487}H^{0.45} \tag{2-5}$$

$$R_{PYS}=3.50d^{0.51}H^{0.24} \tag{2-6}$$

$$R_{PS}=2.35d^{0.62}H^{0.26} \tag{2-7}$$

where R_{PY_1} is the wetted radius of the PY_1 impact sprinkler in m, P_{PYS} is the wetted radius of the PYS plastic impact sprinkler in m, R_{PS} is the wetted radius of the PSH、PSBZ progressive fluidic sprinkler in m, d is the nozzle diameter in mm, and H is the working pressure of the sprinkler in m.

4. Water application rate

(1) Point water application rate

The point water application rate is the ratio of the water depth during the rotating period of a sprinkler in a small area. The formula is shown as the following:

$$\rho_i=\frac{h}{t} \tag{2-8}$$

where ρ_i is the point water application rate in mm/h, h is the water depth in mm, and t is the rotating time of a sprinkler in h. In using the calculation, complete periods of rotations should be utilized.

(2) Average water application rate

The average water application rate is determined by the average amount of each point measured in a controlled water application area. If the areas of each measuring points are the same, the formula of the average water application rate is shown as the following:

$$\bar{\rho}=\frac{\sum_{i=1}^{n}\rho_i}{n} \tag{2-9}$$

where $\bar{\rho}$ is the average water application rate in mm/h, n is the number of points which represent the same size, and ρ_i is the point of each water application rate in mm/h.

If the area of each measuring point is different, the water application rate should be calculated according to the weighted average of the area represented by each measuring point, using the following formula:

$$\bar{\rho} = \frac{\sum_{i=1}^{n} S_i P_i}{\sum_{i=1}^{n} S_i} \tag{2-10}$$

where S_i is the area of the point in m^2, and n is the numeric amount of each point.

5. Water distribution characteristics

The distribution curve of spray water is mainly used to indicate the relationship between the intensity of the sprinkler and the corresponding location of the measurement points, as shown in Figure 2-1.

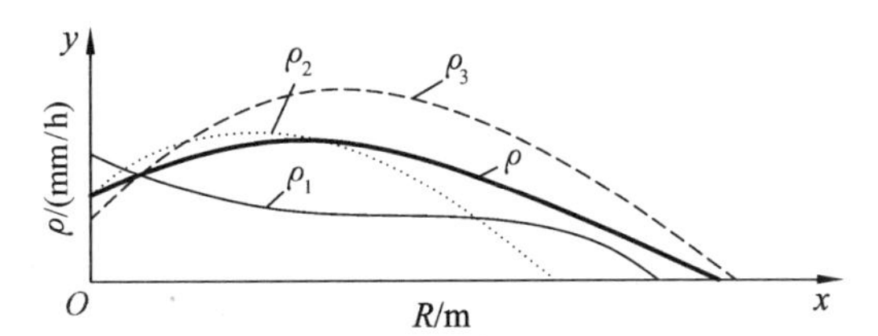

Figure 2-1 Water distribution curves for sprinkler

ρ_1, ρ_2, and ρ_3 are the distribution curves of three sprinklers after the experiment. $\bar{\rho}$ is the average distribution curve of this type of sprinkler.

6. Uniformity of sprinkler irrigation

The uniformity of sprinkler irrigation is the uniformity of the distribution of water in the irrigation area. The practice shows that the uniformity of spray on the whole irrigated area has a decisive effect on crop yield. Therefore, it is one of the technical requirements of agriculture and one of the important indexes to measure the quality of sprinkler irrigation.

J. E. Christiansen first put forward quantitative indexes to evaluate the uniformity of water distribution in sprinkler irrigation. The Christiansen's uniformity coefficient (C_u) is stipulated in ISO 15886-3. The formula is shown as the following:

$$C_u = 100\left(1 - \frac{\sum_{i=1}^{i=n} |X_i - X^-|}{n \times X^-}\right) \tag{2-11}$$

where C_u is the Christiansen's uniformity coefficient in %, n is the total number of collectors, X_i is the water depth collected by collector No. i in mm, and X^- is the average water depth collected in every collector in mm.

Formula (2-11) demonstrates that the more uniformly sprayed irrigation water is, the larger the C_u, though it is always less than 1.0. Under the wind speed of design, the uniformity coefficient of sprinkler irrigation should not be less than 75%, and the uniformity coefficient should not be less than 85% for the traveling sprinkler irrigation system.

7. Droplet impact strength

The droplet impact strength is the impact kinetic energy of droplets on soil or crops in the area of sprinkler irrigation. It depends on factors such as the mass, velocity, and density of a droplet. A droplet's impact strength is commonly characterized by the droplet's diameter or atomization index.

(1) Droplet diameter

The droplet size is one of the main parameters to evaluate the hydraulic performance of a sprinkler. Studies show that when the droplets are too big, the crops are affected more easily by the impact; it is easy for runoff and erosion to form on the soil's surface because the water can't permeate the soil in time. Conversely, when the droplets are too small, although it is beneficial for the growth of crops, the energy consumption of the sprinkler system increases and the sprinkler wetted radius is affected, and the droplets are prone to evaporation and drift. The experiment shows that the optimal range of a droplet diameter at the end of the sprinkler wetted radius is 1 to 3 mm. In each measurement range, the droplet diameter increases gradually with the increase of the nozzle diameter, having a directly proportional relationship. When the working pressure increases, the droplet diameter increases gradually in an inverse proportion. Conversely, the farther away from the sprinkler, the larger the droplet diameter. Also, the droplet size is also affected by the wetted radius and elevation angle of the sprinkler. During experimentation, the methods used for measuring the droplet diameter mainly are the powder method, stain method, optical collector method, laser method, and photographic method.

(2) Atomization index

The atomization index is used to reflect the degree of pulverization and the degree of impact of the jet flow when using a sprinkler or designing a sprinkler irrigation system. The formula is shown as the following:

$$P_{\mathrm{d}}=\frac{H}{d} \tag{2-12}$$

where P_{d} is the atomization index, H is the working pressure of a sprinkler in m, and d is the nozzle diameter in mm.

For a nozzle, the larger P_{d} is, the higher the atomization degree. This creates a smaller droplet diameter and impact.

2.2 Impact sprinkler

2.2.1 Structure of the impact sprinkler

There are many structural types of impact sprinklers. Depending on the number of nozzles, the sprinklers can be categorized as single-nozzle sprinkler, two-way double-nozzle sprinkler, one-way double-nozzle sprinkler, and three-nozzle sprinkler. Each is categorized by its unique spray form, which can be one of two types: either a fan spray sprinkler with a commutation mechanism or a full circle spray sprinkler with no commutation mechanism. Figure 2-2 is the structure of the PY impact sprinkler.

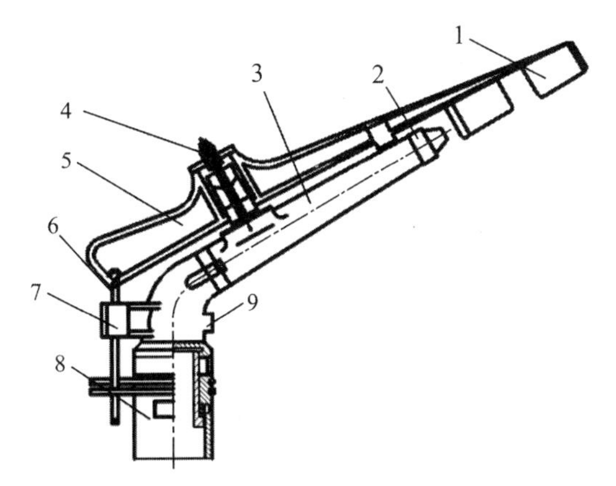

1—Drive vane; 2—Nozzle; 3—Tube & Elbow; 4—Shift shaft; 5—Drive arm; 6—Trip lever; 7—Reverse stops; 8—Lower bearing; 9—Secondary nozzle

Figure 2-2　Structure of the PY impact sprinkler

The basic structure of the impact sprinkler is made up of the following parts:

1. Flow tube

The flow tube is where the water flows through the sprinkler, including the hollow shaft, sprinkler body, tube, regulator, the nozzle, and other parts.

2. Rotary sealing mechanism

The commonly used rotary sealing mechanisms are the radial seal and end face seal, which is made up of parts such as an anti-friction sealing ring, rubber mat (or rubber ring), and sand proof spring.

3. Driving mechanism

The driving mechanism is composed of a drive arm, shift shaft, drive arm spring, and spring seat. Its function is driving the sprinkler to rotate.

4. Reversing mechanism

Also known as the fan mechanism, the reversing mechanism is composed of three parts: the commutator, the limit collar, and the inversion part. Its function is to make the sprinkler make a round trip fan spray in the prescribed angle.

5. Connector

The impact sprinkler and the water supply pipe often has a thread connection; its connector is usually the hollow shaft sleeve of a sprinkler.

The structural difference between an impact sprinkler and other rotating sprinklers is the driving mechanism, which is controlled by a drive arm. The drive arm moves around the axis in the action of the jet stream and hits the sprinkler body with a larger impact impulse, making the sprinkler rotate. This kind of intermittent impact driving torque makes a strong impact within a short time, evening and steadying the sprinkler and directing the jet flow. Thus, the impact sprinkler has a longer wetted radius and a higher uniformity.

The drive arm, drive arm spring, and spring seat is all set on the shift shaft. The shift shaft is fixed to the sprinkler body; most are placed above the tube, and some are placed below the tube. One end of the drive arm spring is inserted into the drive frame and the other end is inserted into the spring seat. The upper end of the spring seat opens many slots, and the rotating spring seat can adjust the spring force of the spring to change the opening angle of the impact. To facilitate the adjustment of the water depth of a fluid director and reduce friction resistance, the impact arm often adopts a suspension structure.

The drive arm is composed of a drive arm body, drive arm bushing, and drive vane. Among them, the drive vane is composed of the deflector plate and the guide plate. The drive arm swings back and forth on the shift shaft under the alternating current of the jet and spring. Its main function is to drive the sprinkler rotation and regulate the water distribution and partial function of the jet.

2.2.2 Operating principle of the impact sprinkler

The drive arm transfers and converts different energies in its working process, which can be divided into five stages.

1. The drive arm detaches from the jet

The initial condition of the drive arm is still. First, the jet flow from the nozzle impacts the deflector plate, which causes the jet to deflect from the axis to the guide plate, and the guide plate obtains the maximum kinetic energy. The guide plate obtains the reaction from the jet, making the drive arm gain kinetic energy and swing outward, rotating around the shift shaft. Then, the drive arm spring twists and produces a torsional moment. Because the torque that the jet exerts on the guide plate is much larger than the torsional moment of the spring, the drive arm has a nappe separation with a fast angular velocity.

2. The deceleration stage of the drive arm

The drive arm is detached from the jet and cut off from the power source, but under the movement of inertia, the drive arm continues to swing in slow motion until it reaches the maximum opening angle, to get the maximum torque. At this point, the angular velocity changes to zero (0), the drive arm stops moving, the maximum potential energy of the spring is reached, and the kinetic energy of the drive arm is converted into elastic potential energy for the spring.

3. The backswing stage of the drive arm

The drive arm spring starts to work after the drive arm stops moving. When the spring creates a torsional movement, the elastic potential energy of the spring is transformed into rotational kinetic energy for the drive arm. The drive arm begins to swing back and the angular velocity increases until the drive arm is about to cut into the jet.

4. The stage of the drive arm reentering the water

At this stage, the drive arm with maximum rotational kinetic energy reenters the jet. The deflector plate of the drive arm is subjected to water first, and then the flow direction is changed. The deflector plate will gain kinetic energy to accelerate the drive arm into the jet.

5. The collision stage of the drive arm

Under the action of the rotary inertial force and tangential side force of the deflector plate, the drive arm collides into the tube with large angular velocity. Because the collision time is short and the impulse moment is very large, the sprinkler can rotate to any given angle by overcoming the friction resistance moment. After the collision, the drive arm completes a full rotation process. The sprinkler stops quickly under frictional resistance and then repeats the rotation process continuously.

For a fan spray sprinkler, the reversal process is relatively simple. The protuberance of the commutator restricts a large swing of the drive arm, which was the collision that occurred at the end of the first stage (the drive arm detaching from the jet). At the back of the collision point, energy is passed between the back of the drive arm and the protruding components of the commutator; the direction of the impulse is changed, so the sprinkler is reversed. The reversal process is rapid, the wetted radius is relatively close, and the droplets are scattered in relatively close quarters.

2.2.3 Models and performance parameters of the impact sprinkler

1. The models of sprinklers

(1) Specification of the models

The sprinklers are made up of Chinese pinyin capital letters and Arabic numerals.

(2) Examples

A double-nozzle impact sprinkler with a 15 mm inlet nominal diameter is marked as 15PY2.

A double-nozzle impeller sprinkler equipped with a commutation mechanism and 50 mm inlet nominal diameter is marked as 50PL2H.

A single-nozzle vertical impact sprinkler with a 40 mm inlet nominal diameter is marked as 40PYC.

A single-nozzle jet sprinkler with a 30 mm inlet nominal diameter is marked as 30PS.

Another single-nozzle sprinkler with a 30 mm inlet nominal diameter is marked as 20P.

2. Basic parameters

The basic parameters of the most commonly used rotary single-nozzle sprinklers are following Table 2-2. The basic parameters of the most commonly used rotary double-nozzle sprinkler are following Table 2-3. For circular nozzles, the diameter refers to the values in Table 2-2 and Table 2-3. For non-circular nozzles, the nozzle diameter is measured by calculating the equivalent diameter, which is determined by the GB/T 19795.1—2005. The flow rate and wetted radius of a sprinkler can be calculated by interpolation when the nozzle diameter is not following the values listed in Table 2-2 and Table 2-3.

Table 2-2 The basic parameters of the rotary single-nozzle sprinkler

Inlet nominal diameter D/mm	Nozzle diameter d/mm	Rated working pressure p/kPa	Sprinkler flow rate Q / (m^3/h)	Sprinkler wetted radius R/m	
				Trajectory angle α	
				20°~25°	26°~30°
10	2.5	150	0.27	9.0	9.5
		200	0.31	9.3	9.8
		250	0.35	9.6	10.2
		300	0.38	10.0	10.5
	3.0	150	0.39	9.5	10.0
		200	0.45	9.8	10.3
		250	0.51	10.1	10.6
		300	0.56	10.5	11.0
	3.5	150	0.58	10.0	10.5
		200	0.62	10.3	10.8
		250	0.69	10.6	11.1
		300	0.75	11.0	11.5
	4.0	200	0.81	10.5	11.5
		250	0.90	10.8	11.8
		300	0.98	11.1	12.1
		350	1.06	11.5	12.5
	4.5	200	1.02	11.0	12.5
		250	1.14	11.5	13.0
		300	1.25	12.0	13.5
		350	1.35	12.5	14.0

Continued

Inlet nominal diameter D/mm	Nozzle diameter d/mm	Rated working pressure p/kPa	Sprinkler flow rate Q / (m^3/h)	Sprinkler wetted radius R/m	
				Trajectory angle α	
				20°～25°	26°～30°
15	4.0	200	0.81	11.5	13.0
		250	0.90	12.0	13.5
		300	0.98	12.5	14.0
		350	1.06	13.0	14.5
		400	1.14	13.5	15.0
	4.5	200	1.02	12.0	13.5
		250	1.14	12.5	14.0
		300	1.25	13.0	14.5
		350	1.35	13.5	15.0
		400	1.44	14.0	15.5
	5.0	200	1.26	13.5	14.5
		250	1.40	14.0	15.0
		300	1.54	14.5	15.5
		350	1.66	15.0	16.2
		400	1.78	15.5	17.0
	5.5	200	1.52	14.0	15.0
		250	1.70	14.8	15.8
		300	1.86	15.5	16.5
		350	2.01	16.0	17.0
		400	2.15	16.5	17.5
	6.0	200	1.81	14.5	15.5
		250	2.02	15.5	16.5
		300	2.22	16.5	17.5
		350	2.40	17.0	18.0
		400	2.56	17.5	18.5
20	6.0	250	2.02	16.0	17.0
		300	2.22	17.0	18.0
		350	2.40	17.5	18.5
		400	2.56	18.0	19.0
		450	2.72	18.5	19.5

Continued

Inlet nominal diameter D/mm	Nozzle diameter d/mm	Rated working pressure p/kPa	Sprinkler flow rate Q/(m^3/h)	Sprinkler wetted radius R/m	
				Trajectory angle α	
				20°~25°	26°~30°
20	6.5	250	2.38	16.5	17.5
		300	2.60	17.5	18.5
		350	2.81	18.0	19.0
		400	3.01	18.5	19.5
		450	3.19	19.0	20.0
	7.0	250	2.76	17.0	18.0
		300	3.02	18.0	19.0
		350	3.26	18.5	19.5
		400	3.49	19.5	20.5
		450	3.70	20.0	21.0
	7.5	250	3.17	17.5	18.5
		300	3.47	18.5	19.5
		350	3.75	19.0	20.0
		400	4.00	20.0	21.0
		450	4.25	20.5	21.5
	8.0	250	3.60	18.0	19.0
		300	3.95	19.0	20.0
		350	4.26	20.0	21.0
		400	4.56	21.0	22.0
		450	4.84	21.5	22.5
	8.5	250	4.27	18.5	19.5
		300	4.46	19.5	20.5
		350	4.81	20.5	21.5
		400	5.15	21.0	22.5
		450	5.46	22.0	23.0
	9.0	250	4.56	19.0	20.0
		300	5.00	21.0	22.0
		350	5.40	21.5	23.0
		400	5.77	22.0	23.5
		450	6.12	22.5	24.0

Continued

Inlet nominal diameter D/mm	Nozzle diameter d/mm	Rated working pressure p/kPa	Sprinkler flow rate Q /(m^3/h)	Sprinkler wetted radius R/m	
				Trajectory angle α	
				20°～25°	26°～30°
30	9.0	300	5.00	21.0	22.5
		350	5.40	22.0	23.5
		400	5.77	22.5	24.0
		450	6.12	23.0	24.5
		500	6.45	23.5	25.0
	9.5	300	5.57	21.5	23.0
		350	6.01	22.5	24.0
		400	6.43	23.0	24.5
		450	6.82	23.5	25.0
		500	7.19	24.0	25.5
	10.0	300	6.17	22.0	23.5
		350	6.66	23.0	24.5
		400	7.12	24.0	25.5
		450	7.56	24.5	26.0
		500	7.97	25.0	26.5
	10.5	300	6.80	22.5	24.0
		350	7.35	23.5	25.0
		400	7.86	24.5	26.0
		450	8.33	25.0	26.5
		500	8.78	25.5	27.0
	11.0	300	7.47	23.0	24.5
		350	8.06	24.0	25.5
		400	8.62	25.5	27.0
		450	9.15	26.5	28.0
		500	9.64	27.5	29.0
	11.5	300	8.16	23.5	25.0
		350	8.81	24.5	26.0
		400	9.42	26.0	27.5
		450	10.00	27.0	28.5
		500	10.54	28.0	29.5

Continued

Inlet nominal diameter D/mm	Nozzle diameter d/mm	Rated working pressure p/kPa	Sprinkler flow rate Q / (m^3/h)	Sprinkler wetted radius R/m	
				Trajectory angle α	
				20°～25°	26°～30°
30	12.0	300	8.89	24.0	25.5
		350	9.60	25.5	27.0
		400	10.26	26.5	28.0
		450	10.88	27.5	29.0
		500	11.47	28.5	30.0
40	12.0	350	9.60	26.0	27.5
		400	10.26	27.0	28.5
		450	10.88	28.0	29.5
		500	11.47	29.0	30.5
		550	12.03	29.5	31.0
	13.0	350	11.27	26.5	28.0
		400	12.04	27.5	29.0
		450	12.77	28.5	30.0
		500	13.47	29.5	31.0
		550	14.12	30.0	31.5
	14.0	350	13.06	28.0	29.5
		400	13.97	29.5	31.0
		450	14.82	30.0	32.0
		500	15.62	31.0	33.0
		550	16.38	32.0	34.0
	15.0	350	15.00	28.5	30.5
		400	16.04	29.5	31.5
		450	17.01	31.0	33.0
		500	17.93	32.0	34.0
		550	18.80	33.0	35.0
	16.0	350	17.07	29.5	31.5
		400	18.25	30.5	32.5
		450	19.35	32.0	34.0
		500	10.40	34.5	36.5
		550	21.40	36.0	38.0

Continued

Inlet nominal diameter D/mm	Nozzle diameter d/mm	Rated working pressure, p/kPa	Sprinkler flow rate Q/(m^3/h)	Sprinkler wetted radius R/m	
				Trajectory angle α	
				20°～25°	26°～30°
50	16.0	400	18.25	32.0	34.0
		450	19.35	34.0	35.5
		500	20.40	35.0	37.0
		550	21.40	36.5	38.5
		600	22.35	37.5	39.5
	17.0	400	20.60	33.5	35.5
		450	21.85	35.0	37.0
		500	23.03	36.5	38.5
		550	24.15	37.5	39.5
		600	25.23	38.5	40.5
	18.0	400	23.09	34.5	36.5
		450	24.49	36.0	38.0
		500	25.82	37.5	39.5
		550	27.08	38.5	40.5
		600	28.28	39.5	41.5
	19.0	400	25.73	35.5	37.5
		450	27.29	37.0	39.0
		500	28.77	38.5	40.5
		550	30.17	39.5	41.5
		600	31.51	40.0	42.5
	20.0	400	28.51	36.5	38.5
		450	30.24	38.0	40.0
		500	31.88	39.5	41.5
		550	33.43	40.0	42.5
		600	34.92	41.0	43.5
60	20.0	500	31.88	40.0	42.5
		550	33.43	41.5	44.0
		600	34.92	43.0	45.5
		650	36.35	44.0	46.5
		700	37.72	45.5	48.0

Continued

Inlet nominal diameter D/mm	Nozzle diameter d/mm	Rated working pressure p/kPa	Sprinkler flow rate Q/(m^3/h)	Sprinkler wetted radius R/m	
				Trajectory angle α	
				20°～25°	26°～30°
60	22.0	500	38.57	41.5	44.0
		550	40.45	43.5	46.0
		600	42.25	45.0	47.5
		650	43.98	46.5	49.0
		700	45.64	48.5	51.0
	24.0	500	45.90	44.0	46.5
		550	48.15	45.5	48.0
		600	50.29	48.0	50.5
		650	52.34	49.0	51.0
		700	54.32	50.0	53.0
	26.0	500	53.87	45.5	48.0
		550	56.50	47.5	50.0
		600	59.02	49.0	52.0
		650	61.43	50.0	53.0
		700	63.75	51.0	54.0
80	26.0	600	59.02	49.0	51.5
		650	61.43	50.0	53.0
		700	63.75	51.5	54.5
		750	65.98	54.0	57.0
		800	68.15	56.0	59.0
	28.0	600	68.45	50.0	53.0
		650	71.24	51.5	54.5
		700	73.93	53.0	56.0
		750	76.52	54.5	57.5
		800	79.04	56.5	59.5
	30.0	600	78.57	51.5	54.5
		650	81.78	53.0	56.0
		700	84.87	54.0	57.0
		750	87.85	55.5	58.5
		800	90.73	57.0	60.0

Continued

Inlet nominal diameter D/mm	Nozzle diameter d/mm	Rated working pressure p/kPa	Sprinkler flow rate Q/(m^3/h)	Sprinkler wetted radius R/m	
				Trajectory angle α	
				20°~25°	26°~30°
80	32.0	600	89.40	53.0	56.0
		650	93.05	55.0	58.5
		700	96.56	57.5	60.5
		750	99.95	58.0	61.5
		800	103.23	60.0	63.5
	34.0	600	100.93	55.0	58.0
		650	105.05	58.0	61.0
		700	109.01	60.5	64.0
		750	112.84	62.5	66.0
		800	116.54	64.5	68.0

Table 2-3 The basic parameters of the rotary-double nozzle sprinkler

Inlet nominal diameter D/mm	Nozzle diameter d_{main}/mm×d_{sub}/mm	Rated working pressure p/kPa	Sprinkler flow rate Q/(m^3/h)	Sprinkler wetted radius R/m	
				Trajectory angle α	
				20°~25°	26°~30°
10	2.5×2.0	150	0.44	9.0	9.5
		200	0.51	9.3	9.8
		250	0.57	9.6	10.2
		300	0.62	10.0	10.5
	3.0×2.0	150	0.56	9.5	10.0
		200	0.65	9.8	10.3
		250	0.73	10.1	10.6
		300	0.80	10.5	11.0
	3.5×2.0	150	0.70	10.0	10.5
		200	0.82	10.3	10.8
		250	0.91	10.6	11.1
		300	0.99	11.0	11.5
	4.0×2.5	150	1.12	10.5	11.5
		200	1.25	10.8	11.8
		250	1.36	11.1	12.1
		300	1.47	11.5	12.5

Continued

Inlet nominal diameter D/mm	Nozzle diameter d_{main}/mm×d_{sub}/mm	Rated working pressure p/kPa	Sprinkler flow rate Q/(m^3/h)	Sprinkler wetted radius R/m	
				Trajectory angle α	
				20°～25°	26°～30°
10	4.5×2.5	150	1.33	11.0	12.5
		200	1.49	11.5	13.0
		250	1.63	12.0	13.5
		300	1.76	13.5	14.0
15	4.0×2.5	200	1.12	11.5	13.0
		250	1.25	12.0	13.5
		300	1.37	12.5	14.0
		350	1.48	13.0	14.5
		400	1.58	13.5	15.0
	4.5×2.5	200	1.33	12.0	13.5
		250	1.49	12.5	14.0
		300	1.63	13.0	14.5
		350	1.76	13.5	15.0
		400	1.89	14.0	15.5
	5.0×3.0	200	1.71	13.5	14.5
		250	1.91	14.0	15.0
		300	2.10	14.5	15.5
		350	2.26	15.0	16.2
		400	2.42	15.5	17.0
	5.5×3.0	200	1.97	14.0	15.0
		250	2.21	14.8	15.8
		300	2.42	15.5	16.5
		350	2.61	16.0	17.0
		400	2.79	16.5	17.5
	6.0×3.0	200	2.26	14.5	15.5
		250	2.53	15.5	16.5
		300	2.77	16.0	17.5
		350	3.00	17.0	18.0
		400	3.20	17.5	18.5

Continued

Inlet nominal diameter D/mm	Nozzle diameter d_{main}/mm × d_{sub}/mm	Rated working pressure p/kPa	Sprinkler flow rate Q/(m^3/h)	Sprinkler wetted radius R/m	
				Trajectory angle α	
				20°~25°	26°~30°
20	6.0×3.0	250	2.53	16.0	17.0
		300	2.77	17.0	18.0
		350	3.00	17.5	18.5
		400	3.20	18.0	19.0
		450	3.40	18.5	19.5
	6.5×3.0	250	2.88	16.5	17.5
		300	3.16	17.5	18.5
		350	3.41	18.0	19.0
		400	3.65	18.5	19.5
		450	3.87	19.0	20.0
	7.0×3.5	250	3.45	17.0	18.0
		300	3.78	18.0	19.0
		350	4.08	18.5	19.5
		400	4.36	19.5	20.5
		450	4.63	20.0	21.0
	7.0×4.0	250	3.66	17.0	18.0
		300	4.01	18.0	19.0
		350	4.33	18.5	19.5
		400	4.63	19.5	20.5
		450	4.91	20.0	21.0
	7.5×3.5	250	3.86	17.5	18.5
		300	4.22	18.5	19.5
		350	4.56	19.0	20.0
		400	4.88	20.0	21.0
		450	5.18	20.5	21.5
	7.5×4.0	250	4.07	17.5	18.5
		300	4.46	18.5	19.5
		350	4.81	19.0	20.0
		400	5.15	20.0	21.0
		450	5.46	20.5	21.5

Continued

Inlet nominal diameter D/mm	Nozzle diameter d_{main}/mm× d_{sub}/mm	Rated working pressure p/kPa	Sprinkler flow rate Q/(m^3/h)	Sprinkler wetted radius R/m	
				Trajectory angle α	
				20°~25°	26°~30°
20	8.0×3.5	250	4.29	18.0	19.0
		300	4.70	19.0	20.0
		350	5.08	20.0	21.0
		400	5.43	21.0	22.0
		450	5.76	21.5	22.5
	8.0×4.0	250	4.50	18.0	19.0
		300	4.93	19.0	20.0
		350	5.33	20.0	21.0
		400	5.70	21.0	22.0
		450	6.04	21.5	22.5
	8.5×4.0	250	4.97	18.5	19.5
		300	5.44	19.5	20.5
		350	5.83	20.5	21.5
		400	6.29	21.0	22.5
		450	6.67	22.0	23.0
	9.0×4.0	250	5.46	19.0	20.0
		300	5.98	21.0	22.0
		350	6.46	21.5	23.0
		400	6.91	22.0	23.5
		450	7.33	22.5	24.0
30	9.0×4.0	300	5.98	21.0	22.5
		350	6.46	22.0	23.5
		400	6.91	22.5	24.0
		450	7.33	23.0	24.5
		500	7.73	23.5	25.0
	9.5×4.0	300	6.55	21.5	23.0
		350	7.08	22.5	24.0
		400	7.57	23.0	24.5
		450	8.03	23.5	25.0
		500	8.46	24.0	25.5

Continued

Inlet nominal diameter D/mm	Nozzle diameter d_{main}/mm×d_{sub}/mm	Rated working pressure p/kPa	Sprinkler flow rate Q/(m^3/h)	Sprinkler wetted radius R/m	
				Trajectory angle α	
				20°～25°	26°～30°
30	10.0×4.0	300	7.16	22.0	23.5
		350	7.73	23.0	24.5
		400	8.27	24.0	25.5
		450	8.77	24.5	26.0
		500	9.24	25.0	26.5
	10.0×4.5	300	7.42	22.0	23.5
		350	8.01	23.0	24.5
		400	8.57	24.0	25.5
		450	9.09	24.5	26.0
		500	9.58	25.0	26.5
	10.0×5.0	300	7.71	22.0	23.5
		350	8.33	23.0	24.5
		400	8.91	24.0	25.5
		450	9.45	24.5	26.0
		500	9.96	25.0	26.5
	10.5×4.5	300	8.05	22.5	24.0
		350	8.70	23.5	25.0
		400	9.30	24.5	26.0
		450	9.86	25.0	26.5
		500	10.40	25.5	27.0
	10.5×5.0	300	8.35	22.5	24.0
		350	9.01	23.5	25.0
		400	9.64	24.5	26.0
		450	10.21	25.0	26.5
		500	10.78	25.5	27.0
	11.0×5.0	300	9.01	23.0	24.5
		350	9.73	24.0	25.5
		400	10.40	25.5	27.0
		450	11.40	26.5	28.0
		500	11.63	27.5	29.0

Continued

Inlet nominal diameter D/mm	Nozzle diameter d_{main}/mm×d_{sub}/mm	Rated working pressure p/kPa	Sprinkler flow rate Q/(m^3/h)	Sprinkler wetted radius R/m	
				Trajectory angle α	
				20°~25°	26°~30°
30	11.5×5.0	300	9.70	23.5	25.0
		350	10.48	24.5	26.0
		400	11.21	26.0	27.5
		450	11.89	27.0	28.5
		500	12.53	28.0	29.5
	12.0×5.0	300	10.43	24.0	25.5
		350	11.27	25.5	27.0
		400	12.04	26.5	28.0
		450	12.77	27.5	29.0
		500	13.47	28.5	30.0
40	12.0×5.0	350	11.27	26.0	27.5
		400	12.04	27.0	28.5
		450	12.77	28.0	29.5
		500	13.47	29.0	30.5
		550	14.12	29.5	31.0
	13.0×5.0	350	12.93	26.5	28.0
		400	13.83	27.5	29.0
		450	14.66	28.5	30.0
		500	15.46	29.5	31.0
		550	16.21	30.0	31.5
	14.0×6.0	350	15.47	28.0	29.5
		400	16.53	29.5	31.0
		450	17.54	30.0	32.0
		500	18.49	31.0	33.0
		550	19.39	32.0	34.0
	15.0×6.0	350	17.40	28.5	30.5
		400	18.60	29.5	31.5
		450	19.73	31.0	33.0
		500	20.80	32.0	34.0
		550	21.81	33.0	35.0

Continued

Inlet nominal diameter D/mm	Nozzle diameter d_{main}/mm×d_{sub}/mm	Rated working pressure p/kPa	Sprinkler flow rate Q/(m^3/h)	Sprinkler wetted radius R/m	
				Trajectory angle α	
				20°～25°	26°～30°
40	16.0×6.0	350	19.47	29.5	31.5
		400	20.81	30.5	32.5
		450	22.07	32.0	34.0
		500	23.27	34.5	36.5
		550	24.40	36.0	38.0
50	16.0×6.0	400	20.81	32.0	34.0
		450	22.07	33.5	35.5
		500	23.27	35.0	37.0
		550	24.40	36.5	38.5
		600	25.49	37.5	39.5
	18.0×6.0	400	25.65	34.5	36.5
		450	27.21	36.0	38.0
		500	28.68	37.5	39.5
		550	30.08	38.5	40.5
		600	31.42	39.5	41.5
	18.0×7.0	400	26.58	34.5	36.5
		450	28.19	36.0	38.0
		500	29.72	37.5	39.5
		550	31.17	38.5	40.5
		600	32.55	39.5	41.5
	20.0×7.0	400	32.00	36.5	38.5
		450	33.94	38.5	40.5
		500	35.78	39.5	41.5
		550	37.72	40.0	42.5
		600	39.19	41.0	43.5

2.3 Fluidic sprinkler

2.3.1 Types and structures of fluidic sprinklers

The fluidic sprinkler is a created Chinese water-saving sprinkler based on the "Coanda effect", which uses the wall effect of flow to change the direction of a jet and obtain the driving force of the rotating sprinkler through the reaction force of the flow. That is to say when the compressed water flows through the fluidic element (installed at the exit of the nozzle), the fluidic element has to not only spray the water but also drive the positive and reverse rotation of the sprinkler. Because all the work is achieved by using the characteristics of the jet itself, it is called a fluidic sprinkler[6].

(1) PSF type feedback fluidic sprinkler

In 1981, the PSF-50 type feedback fluidic sprinkler was successfully developed by the Zhenjiang Institute of Agricultural Machinery (now Jiangsu University) and the Lu Four Machine Repair Shop in Qidong County of Jiangsu Province, using the basic theory of the two-phase wall jet[7]. The structure of the PSF type feedback fluidic sprinkler and the structure of the fluidic element is shown in Figure 2-3 and Figure 2-4.

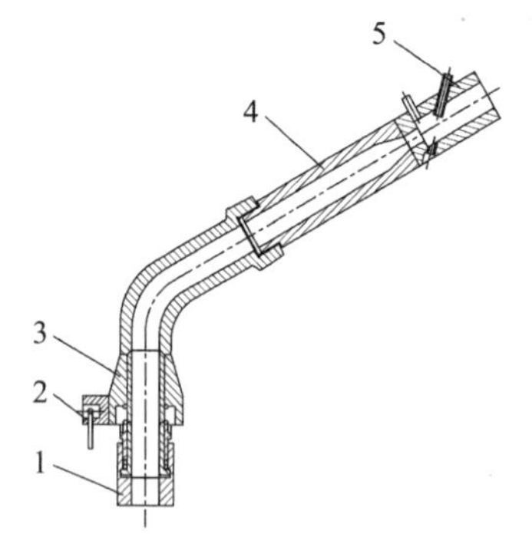

1—Rotating sealing mechanism; 2—Reversing mechanism; 3—Jet body; 4—Nozzle; 5—Fluidic element

Figure 2-3 Structure diagram of the PSF type feedback fluidic sprinkler

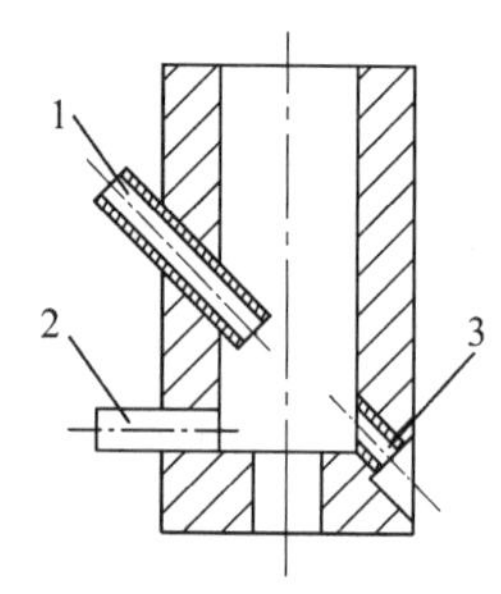

1—Signal water-connected nozzle; 2—Reverse nozzle; 3—Signal water inlet

Figure 2-4 Fluidic element

(2) Continuous fluidic sprinkler

Figure 2-5 shows the three-dimensional section structure of a square cross-section continuous fluidic component. The central line of the channel in the fluidic element is created into a drift angle with the hollow shaft axis in the plumb plane. When the sprinkler rotates forward, the right and left control holes of the fluidic element are opened, and the pressure on the two sides of the interaction zone in the fluidic element is equal. The jet flows along the left wall, producing a counterforce that causes the sprinkler to rotate in the other direction[8].

(3) PSH mutual control stepping fluidic sprinkler

In 1984, the PSH mutual control stepping fluidic sprinkler was designed in Sheng County, Zhejiang province. The structure of the fluidic element of a PSH sprinkler is shown in Figure 2-6.

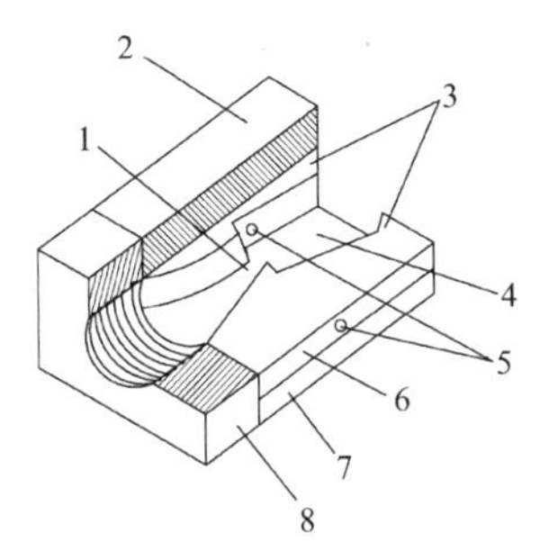

1—Nozzle; 2—Element cover; 3—Element plate; 4—Element interaction zone; 5—Control hole; 6—Element lower cover plate; 7—Rubber pad; 8—Element joint

Figure 2-5 Structure diagram of the fluidic element of square cross-section

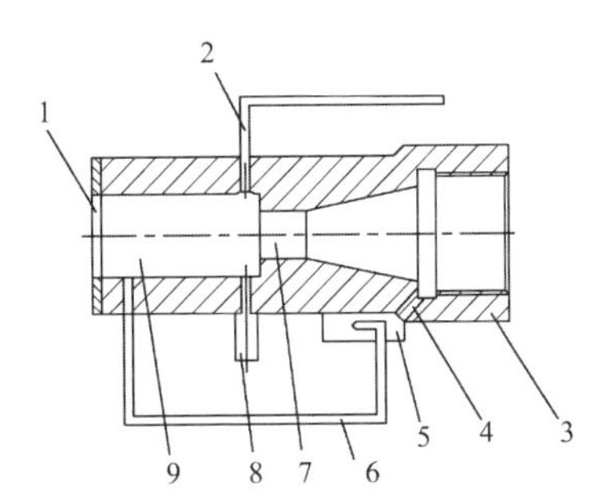

1—Outlet; 2—The plastic tube connecting to the commutator; 3—Main component; 4—Sand-proof ring; 5—Secondary component; 6—Water plastic pipe; 7—Nozzle; 8—Fountain; 9—Interaction zone of the main element

Figure 2-6 Structure diagram of the PSH mutual control stepping fluidic sprinkler

(4) PSZ automatic fluidic sprinkler

Figure 2-7 shows the structure of the PSZ fluidic element[9]. It is a type of a fluidic sprinkler originally developed in Lanxi County, Zhejiang province. The main driving mechanism is an inequipotential fluidic element with a grid feedback flow channel.

(5) Double-click synchronized fluidic sprinkler

In 1990, a type of double-click synchronized fluidic sprinkler was also designed in Sheng County, Zhejiang Province[10]. The structure of a double-click synchronized fluidic sprinkler is shown in Figure 2-8. A deflector plate like the impact sprinkler is installed at the outlet of the fluidic element of this sprinkler. When pressurized water flows from the nozzle, the water clearance controls the signal flow due to the high interaction surface. The back channel blocks the other side of the interaction hole, producing the wall effect under the influence of the atmospheric pressure. At the same time, the refracted water impacts both the jet element and the deflector plate outside the element, thrusting twice to push the sprinkler forward in the function of a double-click synchronized fluidic element. When the instantaneous signal flow is interrupted, the flow returns a direct jet and pushes the sprinkler forward in cycles.

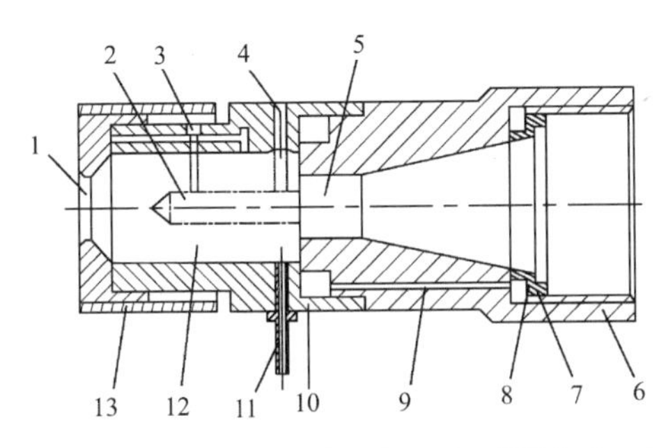

1—Exit; 2—Capacity chamber; 3—Fill hole; 4—Suction negative hole; 5—Nozzle; 6—Element joint; 7—O type sealing ring; 8—Sand-proof ring; 9—Hole of signal source; 10—Wall attachment; 11—Nipple; 12—Interaction zone; 13—Sand shield

Figure 2-7 Structure diagram of the PSZ automatic fluidic sprinkler

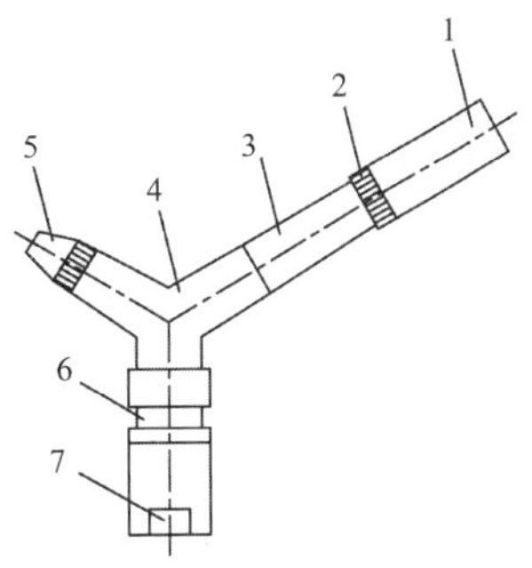

1—Main nozzle; 2—Parallel cap; 3—Straight pipe; 4—Bend body; 5—Side nozzle; 6—Hollow shaft; 7—Pipe nut

Figure 2-8 Structure diagram of the double-click synchronized sprinkler

(6) PXH fluidic sprinkler controlled by clearance

In 2005, a PXH fluidic sprinkler controlled by clearance was designed at Jiangsu University. The assembly diagram of a PXH fluidic sprinkler controlled by clearance is shown in Figure 2-9. In this book, it is referred to as a fluidic sprinkler. Figure 2-10 shows the cross-section of the fluidic element. There are potential differences in the design of a PXH fluidic sprinkler controlled by clearance. When the sprinkler works, the air is supplied through the reverse air holes on the left side. Meanwhile, the air is supplied by the gap on the right. Until the pressure is equal on both sides, the jet is straight, and the sprinkler is still. The signal flow discharges from the signal mouth and it flows into the water hole to block the gap. Due to the pressure on the left higher than the right, the main jet flows to the right side's wall, and the sprinkler moves in step. At this time, the signal mouth is void and connected to the air. After pumping out the signal water from the catheter, the air enters the hole, and the main jet returns to being straight. When the reverse hole is blocked, the pressure on the right side is larger than the left. The main jet is attached to the wall and the sprinkler rotates forward.

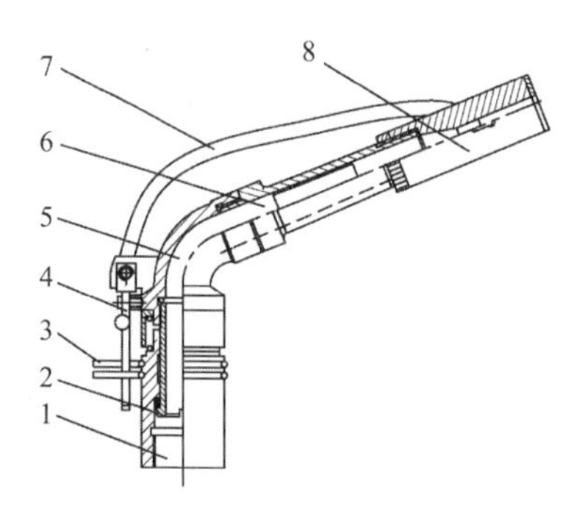

1—Swivel connecting seat; 2—Hollow shaft; 3—Spacing ring; 4—Reversing mechanism; 5—Jet body; 6—Nozzle; 7—Reverse plastic tube; 8—Jet element body

Figure 2-9 Assembly diagram of the complete fluidic sprinkler

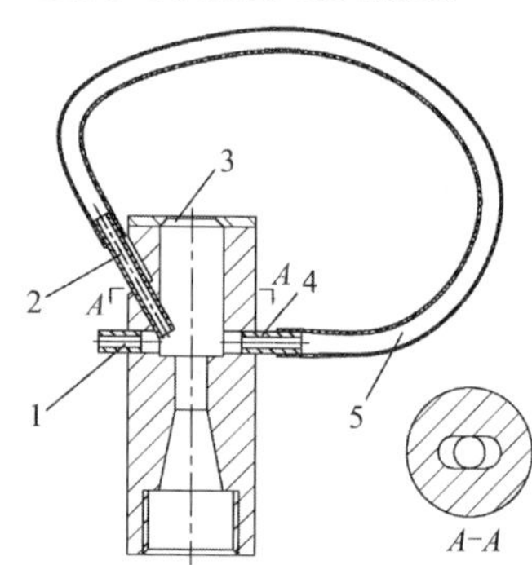

1—Reverse air supply nozzle; 2—Nozzle which receives signal water; 3—Outlet cover plat; 4—Inlet nozzle; 5—Draft tube

Figure 2-10 Fluidic element body

The fluidic element body is composed of a reverse air supply nozzle, a nozzle which receives signal water, an outlet cover plate, a signal water inlet nozzle, and a draft tube. One side of the outer wall of the fluidic element body is provided with a nozzle which receives signal water and a reverse air supply nozzle, and the other side is provided with a signal water inlet nozzle. The nozzle which receives signal water is connected with the inlet nozzle through the draft tube and there is an outlet plate provided on the fluidic element. Using the wall effect, a low-pressure vortex zone is formed intermittently on one side of the main jet due to the interception and flow of the signal water in the nozzle which receives signal water. The opening and closing of the reverse air supply nozzle cause the transformation between the high pressure and low pressure. This occurs so that the main jet forming the pressure difference between the two ends of the wall can let the water flow, completing the directing, stepping and reversing operation.

2.3.2 Design theories of fluidic sprinklers

1. Working principle and technical parameters of fluidic sprinklers

The following is the operational principle of fluidic sprinklers:

Direct state: As shown in Figure 2-11a, the water flows from hole D into the action zone, and the main jet is ejected from the central circular hole. The left and the right sides of the main jet are separated from each other, and the air at both ends no longer circulates. By opening the reverse air mouth of the left water jet element, gas is added to the left cavity. Sir gap C between the exit at the right side of the element and the water jet is fed into the air, so the pressures on both sides of the main jet are equal, the main jet is in direct current, and the nozzle is still. At the same time, the nozzle which receives signal water on the left edge of the water jet, then the signal water flows to the inlet nozzle in the draft tube. The length of I in Figure 2-11a indicates the depth of the signal nozzle.

Step state: As shown in Figure 2-11b, the water from the nozzle which receives signal water flows into the inlet nozzle, gap C continuously shrinks and eventually becomes blocked, forming a low-pressure vortex area on the right. So, the pressure on the left grows larger than the right. When the pressure difference reaches a certain value, the main jet flows to the right side's attachment wall, and the water flows through the chamfering of the outlet cover plate to generate a propelling force to the sprinkler, rotating to the right. Under the wall-attached condition, the nozzle is void and no signal water is received due to the main jet bending. Following the release of the signal water from the draft pipe, air enters the water inlet through the draft tube and the pressure of both sides is equalized. The main jet finishes its working process. The cycle repeats and the sprinkler is automatically controlled, thus producing direct-step-direct-cyclic action.

Reverse state: As shown in Figure 2-11c, the reverse tube is connected with the reversing mechanism. Adjusting the spacing ring can control the spray angle of the sprinkler. When the sprinkler converts to its reversing mechanism, which will be restrained by the limit ring, the reverse air supply nozzle is blocked, leaving the left chamber airless, and forming a low-pressure vortex area. However, the right chamber is still supplied with air from gap C, and pressure on the right is larger than that on the left. With the main flow to the left wall still attached, the sprinkler continues to reverse rotation. Rotating to the other side of the limit ring, the reverse air supply opens again, the air enters the left cavity and is balanced with the pressure difference of the right cavity. Then, the reverse operation is stopped until the direct return is resumed. In this continuous cycle, the fluidic sprinkler completes the direct—step—reverse—direct—step spray operation.

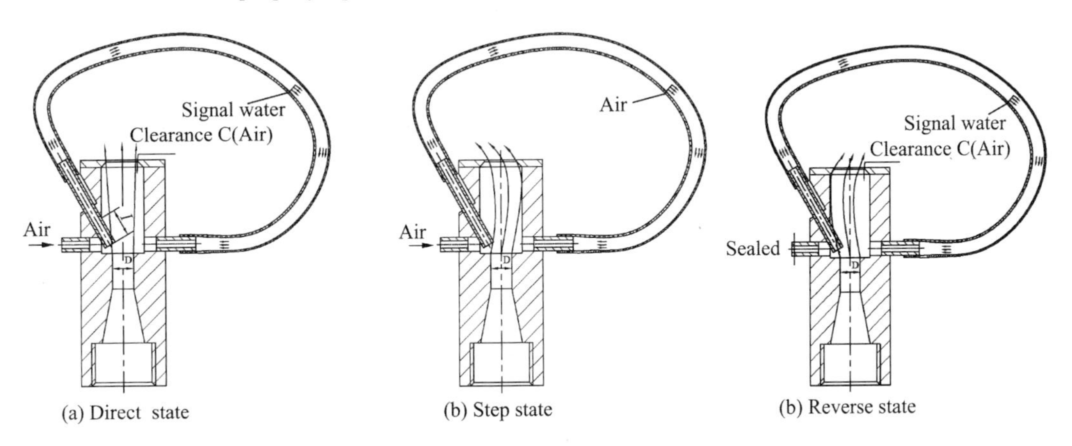

Figure 2-11 Schematic diagram of three states

Because the fluidic element body effectively and conveniently controls the interrupted interception of the water in the water interface, the limit ring can effectively and conveniently control the opening or closing of the reverse nozzle, creating automatic intermittent work. Table 2-4 is the main technical parameters of fluidic sprinklers. Research shows that when the performance of the parameters meets the requirements of Table 2-4, the spray uniformity, droplet sizes, and atomization indexes are superior to those of the PY2 metal sprinkler. The price of the PXH fluidic sprinkler controlled by clearance is reduced by more than 20% than the current agricultural sprinkler, filling the blank of low pressure sprinklers in China.

Table 2-4 Main technical parameters of the fluidic sprinkler

Type	Nozzle diameter/mm	Working pressure/kPa	Flow rate/(m^3/h)	Wetted radius/m	Water application rate/(mm/h)
10PXH	4	250	1.00	12	2.10
15PXH	6	300	1.70	16	2.20
20PXH	8	350	3.50	20	2.50
30PXH	10	400	7.50	26	3.60
40PXH	14	450	15.60	33	4.60
50PXH	18	500	27.00	40	5.30

2. Important structural parameters

Figure 2-12 and Figure 2-13 show the three-dimensional drawing and the schematic diagram of the fluidic element. In a fluidic element, the important structural parameters should include the shrinkage angle γ, a base and a fill hole, an inlet hole, the angle of the water intake hole β, potential difference H (mm), and the length of the zone of action L (mm). Line A in the figure indicates the position at the export base circle.

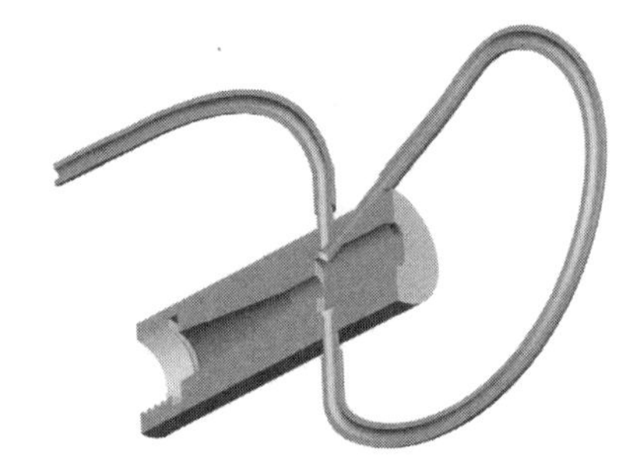

Figure 2-12 3D drawing of the fluidic element

Determination of the shrinkage angle γ of the jet element[11-13]: to ensure that the flow has a smaller degree of turbulence in the nozzle outlet, the velocity distribution at the nozzle outlet pipe should be consistent with the flow velocity distribution at the turbulent state. That is to say, the maximum velocity along the flow direction should be located in the tube axis. So the inlet section of the nozzle should have a smooth transition section and a straight line at the outlet end so that the flow can be adjusted, stabilized, and contracted to achieve a stable effect. In the design process, too large of a contraction angle will lead to insufficient flow stability, and too small will increase the overall size of the sprinkler.

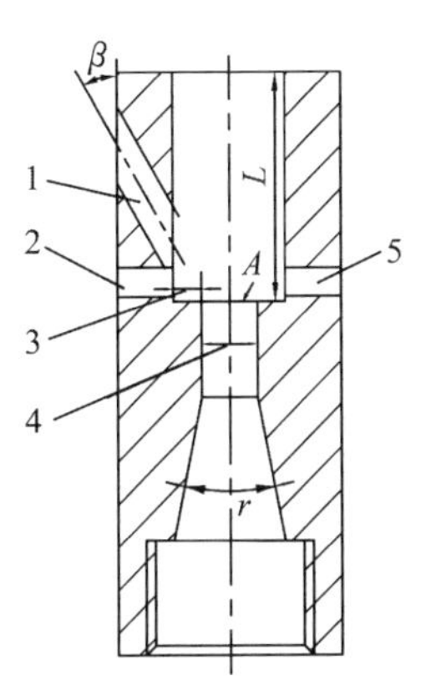

1—Water intake hole; 2—Fill hole; 3—Potential difference; 4—Base hole; 5—Inlet hole

Figure 2-13 Fluidic element

Determination of the diameter of the base hole D(mm): it is always important to determine the nozzle size. The sectional area of a nozzle can be calculated according to the following formula:

$$A=\frac{Q}{\mu\sqrt{2gH}} \tag{2-13}$$

where A is the sectional area of the nozzle in m^2, Q is the sprinkler flow rate in m^3/s, μ is the flow rate coefficient, g is the gravity acceleration in m/s^2, and H is the pressure head in m.

Determination of the fill hole and the inlet hole: the fill hole and the inlet hole of the fluidic sprinkler play a guiding role, so the design method is the same.

Determining the angle of the water intake hole: the control part of the fluidic element must ensure the signal flow. It determines the quality of the signal flow at the same time because the main surface is a mixture of gas and water. The gas in the signal flow has a certain influence on the main jet attaching to the wall, so the design of the angle of the water intake hole is important to ensure that the water signal receives more water and less gas.

Determination of the potential difference: the choice of potential difference is to ensure that water weather can be attached to the wall and can produce enough torque to the head forward. This determines the wall performance and thrust size of the element, to ensure that the wall position is as short as possible and the potential difference is not too large.

Determination of the length of the zone of action: The length of the zone of action determines the length of the submerged jet and the amount of gas mixing. The greater the length of submerged jet aeration and disturbance of water flow, the more momentum exchanged in the boundary layer and the greater the resistance loss of water flow. To have the best attachment performance of a jet and reduce the loss of resistance, the zone of action should not be too long.

2.3.3 Structure optimization and design method of the fluidic sprinkler

1. Experimental research

The prototype of the fluidic sprinkler is tested in the indoor irrigation test hall of Jiangsu University. Figure 2-14 shows a photo of the sprinkler test hall, which is a circular hall with a diameter of 44 meters. As an indoor test hall, wind and other factors are excluded. During testing, an RS485 automatic measuring system was used to test the water application rate. A stopwatch with a precision of 0.01 s was used to measure the rotation speed of the sprinkler, and a meter ruler was used to calculate the wetted radius of a sprinkler; this ensures that data is accurate and reliable[14].

Figure 2-14 Sprinkler test hall

During the operation of rotating sprinklers, the step frequency and step angle are important performance indexes of the sprinkler stability. Thus, it is necessary to study the influence of structural parameters of fluidic sprinklers on the step frequency and step angle, and to get the primary and secondary order of each factor.

The Orthogonal test[15, 16] is used to compare the parameters of each structure. The influence of the structural parameters on the step frequency and step angle are analyzed, and the optimum structure sizes of the sprinkler are put forward.

(1) Purposes of the test

The influence of the structural parameters of the sprinkler on the step frequency and step angle is explored, and the optimum structure sizes of the fluidic sprinkler are proposed.

(2) Test factors and test scheme

In the important structural parameters of a fluidic sprinkler, the base hole diameter D is determined by the sprinkler type. The fill hole and the inlet hole play the role of guiding gas and water. Therefore, selecting the step frequency and step angle parameter directly impacts experimental factors such as the potential difference H, the length of the zone of action L, the shrinkage angle γ, and the angle of the water intake hole β. According to the design principle of a sprinkler, the factors H, L, γ and β are represented by A, B, C, and D respectively. Select the 10PXH type sprinkler tested under the working pressure of 250 kPa and the flow rate of 0.82 m^3/h. The table of factor level selection is shown in Table 2-5, using the $L_9(3^4)$ orthogonal table. Table 2-6 is the test influence rule of structural parameters on the step frequency and step angle.

Table 2-5 Factor level table

Level	Factor			
	A	B	C	D
1	2.6	18	14	30
2	2.8	20	20	45
3	3.0	22	26	60

Table 2-6 Test scheme

Test number	A	B	C	D
1	1	1	1	1
2	1	2	2	2
3	1	3	3	3
4	2	1	2	3
5	2	2	3	1
6	2	3	1	2
7	3	1	3	2
8	3	2	1	3
9	3	3	2	1

Figure 2-15 is a photo of fluidic elements in an orthogonal experiment. In order to improve the manufacturing precision during the prototype machining process, the fluidic element is divided into upper and lower sections using a threaded joint, which was adopted by line cutting at the hole of the fluidic element. The machining tolerance and design tolerance were controlled in less than 0.02 mm.

Figure 2-15 Entities of fluidic elements

(3) Analysis of orthogonal test results

Determining the performance index of the sprinkler includes step frequency and step angle. The reciprocal of the sum of the direct time and the wall time of a fluidic sprinkler is the step frequency: $f=\dfrac{1}{t_1+t_2}$. The moment of inertia of the rotating axis of the jet wall is as follows:

$$J\frac{\mathrm{d}\omega}{\mathrm{d}t}=\rho Qru\sin\beta \tag{2-14}$$

where r is the distance from the wall to the shaft in m, u is the average outlet velocity in m/s, and β is the angle between the center line and the wall of the wall in (°)。

When $t=0,\omega=0$.

Upper solution:

$$\omega=\frac{1}{J}\rho Qrut\sin\beta \tag{2-15}$$

Corner $\varphi=\int_0^{t_2}\omega\mathrm{d}t$, the angle of rotation is obtained:

$$\varphi=\frac{1}{2J}\rho Qrut_2^2\sin\beta \tag{2-16}$$

The test result is shown in Table 2-7. It is evident that the six tests of the sprinkler ran reliably, the step frequency was close to 1 Hz, and the step angle was more than 1 degree. The tests of levels 1, 6 and 9 are ideal. The test level $A_2B_3C_1D_2$ used in test six had superior results, with the step frequency being 0.95 Hz and the step angle being the maximum, 2.8 degrees.

Table 2-7 Test results

Test number	Step frequency f/Hz	Step angle φ/(°)
1	1.36	1.8
2	1.72	0.7
3	2.20	0.3
4	0.28	1.7
5	0.40	2.1
6	0.95	2.8
7	1.70	0.1
8	1.51	1.8
9	1.30	2.4

The direct analysis method is used to analyze the test results. When the frequency of each step is subtracted by 1, the absolute value is obtained. Table 2-7 is the calculation result of this test. Table 2-8 shows the relationships between the levels of each factor and the indicators, as shown in Figure 2-16 and Figure 2-17.

Table 2-8 Analysis of test results

Performance index		A	B	C	D
Step frequency f/Hz	K_1	2.28	1.78	0.92	1.26
	K_2	1.37	1.83	1.74	1.47
	K_3	1.51	1.55	2.50	2.43
	$\overline{K_1}$	0.76	0.59	0.30	0.42
	$\overline{K_2}$	0.46	0.61	0.58	0.49
	$\overline{K_3}$	0.50	0.52	0.83	0.81
	R	0.30	0.09	0.53	0.39
Step angle φ/(°)	K_1	2.80	3.6	5.30	6.30
	K_2	6.60	4.6	4.80	3.60
	K_3	4.30	5.5	2.50	3.80
	$\overline{K_1}$	0.93	1.2	1.77	2.10
	$\overline{K_2}$	2.20	1.53	1.60	1.20
	$\overline{K_3}$	1.43	1.83	0.83	1.27
	R	1.27	0.63	0.93	0.90

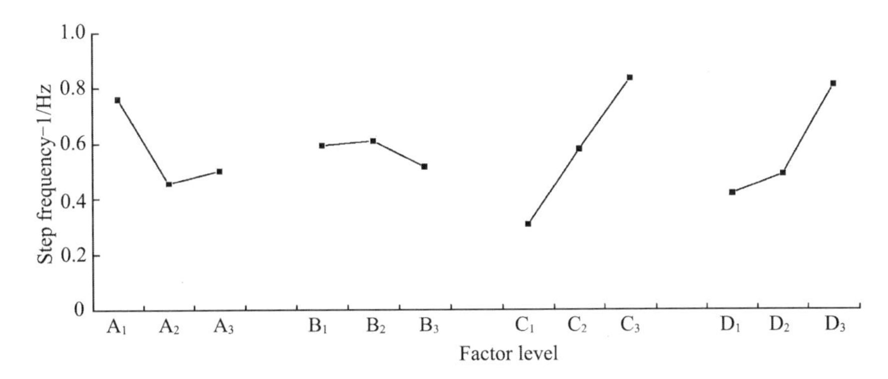

Figure 2-16 Relationship between step frequency and factor level

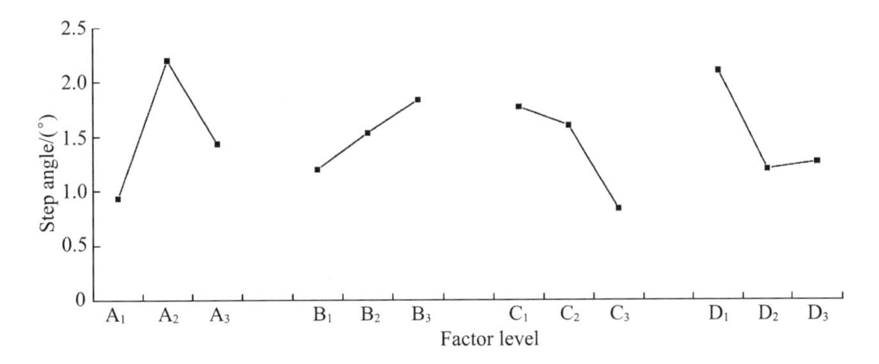

Figure 2-17 Relationship between step angle and factor level

From the magnitude of the range R in Table 2-8, it is clear that the main sequence of each factor affecting the step frequency f is CDAB, and the main order of the factors affecting the step frequency is ACDB.

The following conclusions can be drawn from Table 2-8, Figure 2-16 and Figure 2-17:

Factor A: potential difference H varied from 2.6 to 3 mm. When the potential difference is 2.8 mm, the fluidic element is in the best working condition, step frequency is close to 1 Hz, and the step angle is the maximum. In addition, when the diameter of the base hole D is 4 mm, the potential difference is less than 2.6 mm and greater than 3 mm. Changing any other parameter would lead to the failure of the nozzle.

Factor B: the length of the zone of action L varied from 18~22 mm. The larger the action zone is, the closer the step frequency is to 1 Hz, and the larger the step angle is. Also, when the diameter of the base hole D is 4 mm, the action area is less than 18 mm and greater than 22 mm. Changing any of these parameters would also be fatal to the working capabilities of the nozzle.

Factor C: the range of the shrinkage angle is between 14° to 26°; the greater the contraction angle, the more gradually the step frequency deviates from 1 Hz. The smaller the step angle is, the optimum angle of contraction is 14°.

Factor D: the range of the water intake hole angle is between 30° to 60°, and the greater the angle of water intake hole is, the more gradually the step frequency deviates from 1 Hz. The smaller the step angle is, the optimum angle of the water intake hole is 30°.

2. Design method of a fluidic sprinkler[17]

Figure 2-18 Series of sprinklers

It is of great academic value and practical significance to put forward the theory and design method of the fluidic sprinkler. The same orthogonal experiments were carried out on 15PXH, 20PXH, 30PXH, 40PXH, and 50PXH sprinklers. Figure 2-18 is a photo of sprinklers in orthogonal experiments; each type of manufacturing sprinkler has 9 prototypes. A total of 45 tests were carried out. Some important structural parameters of fluidic sprinklers are listed below.

(1) Calculation formula of the potential difference and the length of the zone of action

① Coefficient formula. Based on a multitude of experimental research data, the calculation formula of the action area length and potential difference of the fluidic element as shown below.

Action area length

$$L=(2.1\sim5.5)D \tag{2-17}$$

Potential deference

$$H=(0.18\sim0.63)D \tag{2-18}$$

where D is the diameter of the base hole in mm. The larger D is, the smaller the coefficient; the coefficient values are shown in Table 2-9.

Table 2-9 Interaction lengths and difference coefficients

Sprinkler type	10	15	20	30	40	50
D/mm	$\phi 4$	$\phi 6$	$\phi 8$	$\phi 10$	$\phi 14$	$\phi 18$
Working pressure/MPa	0.25	0.30	0.35	0.40	0.45	0.50
L/mm	22	24	26	28	34	38
Length coefficient	5.5	4.0	3.5	3.0	2.4	2.1
H/mm	2.5	2.8	3.2	3.7	3.0	3.0
Difference coefficient	0.63	0.43	0.34	0.28	0.21	0.18

② Regression formula.Figure 2-19 is a curve graph of the relationship between the difference ratio (H/D) and the diameter of the base hole. Figure 2-20 is a curve graph of the relationship between the length of the zone of action and the diameter of the base hole. After the regression between Figure 2-19 and Figure 2-20, the functional relationship between the ratio of the different ratio and the diameter of the base hole is as follows:

$$H/D=0.002\ 7D^2-0.094\ 3D+1.007\ 2 \tag{2-19}$$

The functional relation between the length of the zone of action and the diameter of the base hole is as shown below:

$$L = -0.0454D^2 + 2.1501D + 13.7285 \tag{2-20}$$

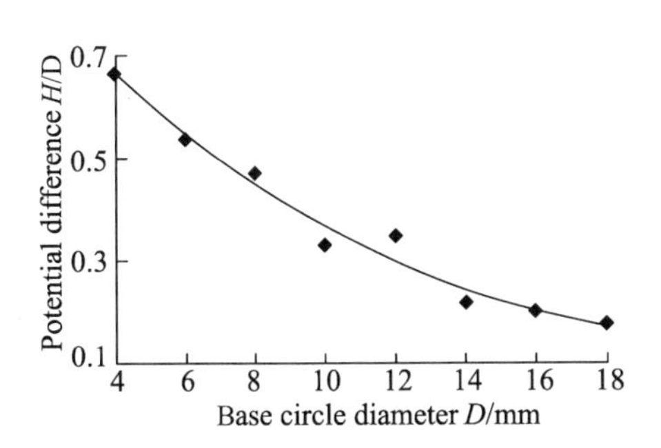

Figure 2-19 Circle curve of potential difference ratio and base circle

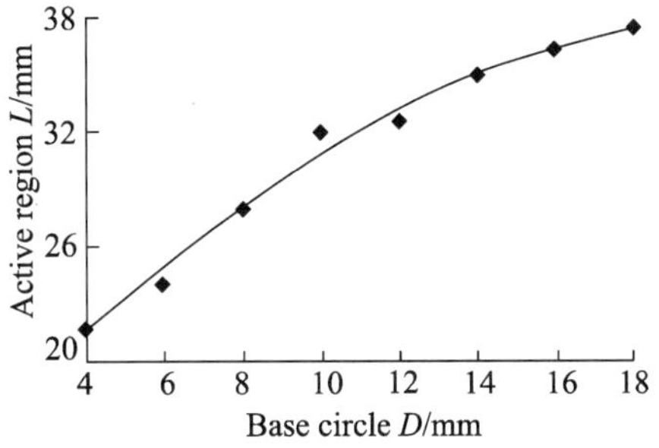

Figure 2-20 Circle of the relationship between the length of the base circle

(2) Size of the outlet cover plate

The size of the outlet cover plate of a fluidic sprinkler is also a key dimension in the sprinkler design, as it directly impacts the normal operation of a sprinkler. The size of the outlet cover plate is comprised of a left potential difference, right potential difference, inner chamfer, and outlet line.

① The recommended formula of the left potential difference:

$$m = (0.08 \sim 0.38)D \tag{2-21}$$

where D is the diameter of the base hole in mm. The larger D is, the smaller the coefficient.

② The recommended formula of the right potential difference:

$$m + c = (0.12 \sim 0.53)D \tag{2-22}$$

The larger D is, the smaller the coefficient. Coefficient values are shown in Table 2-10.

Table 2-10 Potential difference coefficients of the outlet cover plate

Sprinkler model	10	15	20	30	40	50
D /mm	$\phi 4$	$\phi 6$	$\phi 8$	$\phi 10$	$\phi 14$	$\phi 18$
Working pressure/MPa	0.25	0.30	0.35	0.40	0.45	0.50
Left potential difference coefficient	0.38	0.23	0.21	0.21	0.08	0.08
Right potential difference coefficient	0.53	0.28	0.24	0.24	0.12	0.12

③ Length of the outlet line of the cover plate. The length is generally controlled to be between 0.5 to 0.8 mm. If the length is too small, the cover wears down easily, impacting the life of a fluidic element. If the length is too large, the air resistance of the air gap is too big during the air make-up, affecting the range.

④ Left and right inner chamfer of the cover plate. The size of the inner chamfer directly affects the bending direction of the jet step and reverse state, therefore it also influences the driving movement of the step and reverse.

(3) The relation between the length of the tube and the step angle

Through many tests, the relationship between the length of the tube and the step angle for PXH30 sprinkler is found using this formula:

$$\varphi=40l^2-35l+9 \quad (2\text{-}23)$$

where φ is the step angle in (°), and l is the length of the tube in m.

2.3.4 Operation test of the fluidic sprinkler

1. Abrasion test[17]

Because of the working principle and structure characteristics of a fluidic sprinkler, the size of a fluidic element needs to be precise so that the machining accuracy is high. In actual operation, the long-term flow erosion, especially the abrasion of sediment media, will change the size of a fluidic element. If the abrasion of the fluidic element exceeds the allowable operating range, it will lead to the sprinkler's inability to step and rotate, affecting the normal operation and reliability of the sprinkler. Therefore, it is very important to study the abrasion test of the fluidic sprinkler.

(1) Analysis of the worn parts of a sprinkler

From the working principle of the sprinkler, it is evident that the abrasion of the fluidic element mainly affects the following aspects: (ⅰ) contraction taper pipe and base circle section; (ⅱ) left and right side walls of the zone of action; (ⅲ) inner chamfer and outlet line; (ⅳ) signal water intake nozzle. Abrasion usually comes as a result of active surface wear onto the jet working on contraction taper pipes, base circle sections, and the left and right walls of the zone of action. The key concern is that the abrasion of the nozzle directly affects the main jet's ability to intake a sufficient amount of water, the formation of the pressure difference between the left and right sides of the sprinkler, the attachment of the walls to the jet, and its entire normal operation.

The abrasion of the water medium, especially the sand medium on the nozzle size, becomes the key to the stability of the sprinkler. The smaller the abrasion of the nozzle, the more stable the operation. The abrasion of the nozzle size needs to be controlled within the permissible size range. It is possible to reduce the abrasion of the nozzle by considering the following two methods.

① Indirect measures. The nozzle is designed to allow adjustment to its structure. If after a while a sprinkler is not functioning normally, increase the insertion depth of the nozzle, and then the sprinkler can continue to work properly, prolonging the service life of the sprinkler.

② Direct measures. Use abrasion-resistant materials to produce nozzles and other fluidic elements to reduce abrasion.

(2) Abrasion-resistant test device

In order to study the abrasion of a fluidic sprinkler and its influence on the operation of the sprinkler after being worn, the abrasion-resistant device of the self-circulating sprinkler is designed. The test device is shown in Figure 2-21, which consists of a water tank, a water pump unit, a test sprinkler, a flow meter, a pressure gauge, a ball valve, a pipeline, and auxiliary support. The water tank is conical, the top is provided with a cover, and the upper part of the sidewall is provided with a strip opening so that the test nozzle can be extended into the water tank.

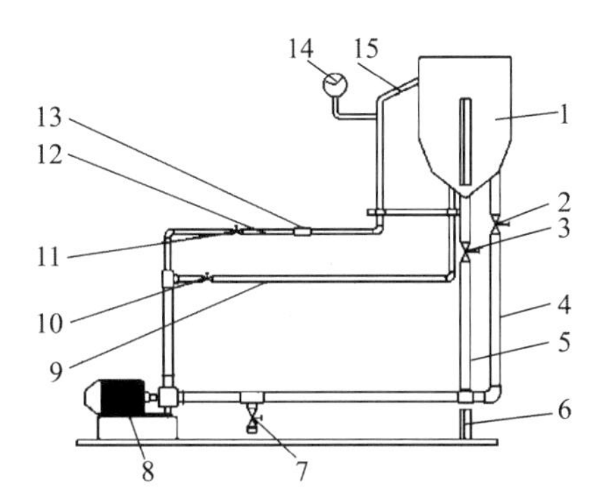

1—Water tank; 2, 3, 7, 10, 11—Ball valve; 4—Water intake pipeline; 5—Drain line; 6—Support; 8—Pump unit; 9—Self mixing line; 12—Sprinkler test pipeline; 13—Flow meter; 14—Pressure gauge; 15—Test sprinkler

Figure 2-21 Schematic diagram of the test device for sprinkler abrasion- resistance

The working process begins by closing the ball valve on the water inlet pipe. The water pump is set by opening the ball valve on the self-mixing pipeline and then closing the ball valve on the sprinkler test pipeline. After a sediment-laden water in the tank is in homogenization, the ball valve on the sprinkler test pipeline can be opened and adjusted so that the inlet pressure gauge of the sprinkler is set to the working pressure. The flow rate is measured by the flowmeter. The sprinkler has a reverse mechanism, making the sprinkler spray at a specified angle. The water flows into the water tank from the opening side of the water tank and runs in circulation. After a necessary amount of time has passed, the ball valve on the sprinkler test pipeline as well as the one on the self-mixing pipeline, the water pump, and the inlet water valve can be closed. If the sand water needs to be drained, the ball valve on the drainpipe can be reopened to drain the sand water and clean the device.

2. Endurance test[17]

In the sprinkler test, the durability of the sprinkler is one of the most important aspects of its hydraulic performance, especially in the actual application of the sprinkler.

In order to test the durability of a fluidic sprinkler, a durability circulation test-bed has been developed. The test-bed consists of a container, pressure gauge, submersible pump, pipeline, and test sprinkler, as shown in Figure 2-22.

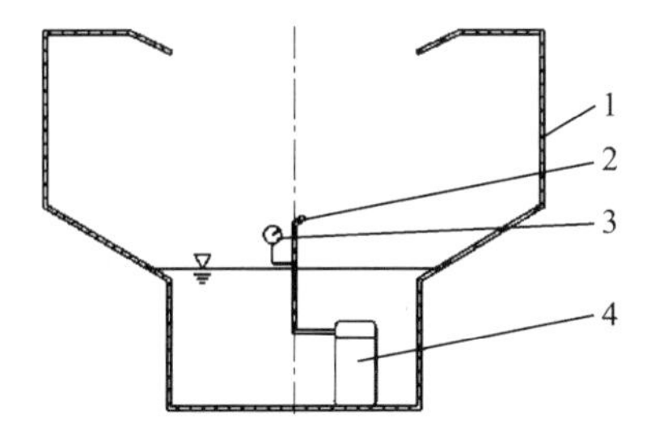

1—Container; 2—Test sprinkler; 3—Pressure meter; 4—Submersible pump

Figure 2-22 Mini sprinkler durability testbed

2.4 Variable-rate sprinkler

2.4.1 Variable-rate spraying theory and its realization method

1. Working parameter relation equations of variable-rate sprinklers

The traditional rotating sprinkler sprays in a full circle, as shown in Figure 2-23.

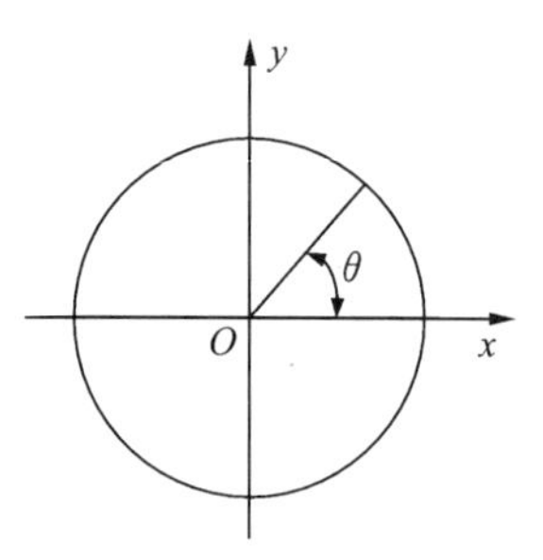

Figure 2-23 Schematic diagram of the full circle spraying

O represents the position of the sprinkler. When the sprinkler rotates from the x-axis to the y-axis, the spray area of the sprinkler can be calculated as follows[18]:

$$S=\pi R^2 \cdot \frac{\theta}{2\pi}=\frac{1}{2}R^2\theta=\frac{1}{2}R^2\omega t \tag{2-24}$$

where R is the wetted radius of the sprinkler in m, ω is the rotational angular velocity of the sprinkler in r/s, t is the rotation time of the sprinkler in s, and θ is the spray angle in (°).

Substituting Formula (2-24) with Formula (2-25) calculates the average water application rate and the relationship between the flow rate and the wetted radius.

$$\rho=\frac{\bar{h}}{t}=\frac{1\,000Q}{\frac{1}{2}R^2\omega t} \tag{2-25}$$

where $\bar{h}$ is the average depth of the sprinkler irrigation in mm.

$$Q=\frac{1}{2\,000}\bar{h}\omega R^2 \tag{2-26}$$

In each sprinkler irrigation system, there is a certain water quantity depending on the needs of the crop, so $\bar{h}$ is set as a fixed value. Within the operating range, the change of the flow rate has little influence on the rotation speed. The relation between the flow rate and the wetted radius is obtained using the above formula. The formula is $Q_1/Q_2=R_1^2/R_2^2$. Where Q_1, Q_2 and R_1, R_2 are the flow rate and wetted radius values in two different cases respectively.

2. Theoretical analysis of variable spraying

(1) Square and triangular spray boundary equations

In the application of sprinkler irrigation, there are many layouts of sprinklers [19]. According to the layout of a pipeline and the control area of a sprinkler, the spray area of a single sprinkler is square and triangle, which is beneficial for improving the water utilization and uniformity. The phenomenon of heavy spray, super spray, and spray out will be significantly reduced. When the sprinkler is arranged in a square and triangular area, the control area of the square and the triangle domain is less than the complete circular spraying. However, the overlap rate and super spray rate decrease in terms of the perspective of precision irrigation.

The shape of a square spray area is shown in Figure 2-24. O, located in the center most position of the square, is the sprinkler, and $OA = R_0 = 1$, $OB = R_0$. In the course of rotation, the wetted radius of the sprinkler varies in its four different cycles. From OA to OB, the angle is α, the wetted radius is gradually decreased, and four peaks are generated in a circle. At the maximum, the wetted radius varies rapidly, and the range varies slowly near the minimum.

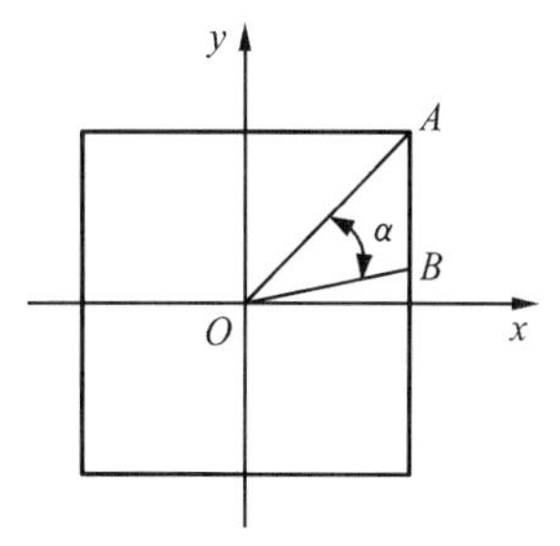

Figure 2-24 Schematic diagram of the square spray

The boundary function of the square can be obtained by Figure 2-24:

$$R=\begin{cases}\dfrac{R_0}{\sqrt{2}\cos(\pi/4-\alpha)} & 0\leqslant\alpha\leqslant\dfrac{\pi}{2}\\ \dfrac{R_0}{\sqrt{2}\cos(3\pi/4-\alpha)} & \dfrac{\pi}{2}\leqslant\alpha\leqslant\pi\\ \dfrac{R_0}{\sqrt{2}\cos(5\pi/4-\alpha)} & \pi\leqslant\alpha\leqslant\dfrac{3\pi}{2}\\ \dfrac{R_0}{\sqrt{2}\cos(7\pi/4-\alpha)} & \dfrac{3\pi}{2}\leqslant\alpha\leqslant 2\pi\end{cases} \tag{2-27}$$

Within a circle, the boundary of the square region is the solution of the equation. The shape of the triangular spray area is shown in Figure 2-25. The sprinkler is in the center O position of triangle, appearing in three peaks, and the variational range of the wetted radius of triangle spray is larger than that of the square.

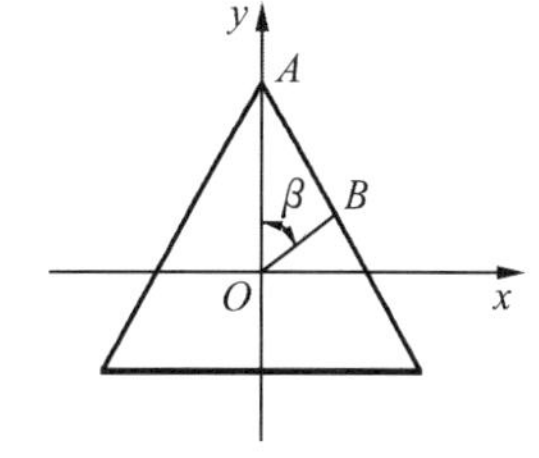

Figure 2-25 Schematic diagram of triangle spray

The triangular boundary function can be obtained by Figure 2-25:

$$R=\begin{cases}\dfrac{R_0}{2\cos(\pi/3-\beta)} & 0\leqslant\beta\leqslant\dfrac{2\pi}{3}\\ \dfrac{R_0}{2\cos(\pi-\beta)} & \dfrac{2\pi}{3}\leqslant\beta\leqslant\dfrac{4\pi}{3}\\ \dfrac{R_0}{2\cos(5\pi/3-\beta)} & \dfrac{4\pi}{3}\leqslant\beta\leqslant 2\pi\end{cases} \tag{2-28}$$

The boundary of the triangle region is the solution to the equation.

(2) Relationship between the flow rate and the wetted radius

There is a direct relationship between the flow rate and the wetted radius: the greater the flow rate, the greater the wetted radius. The change of the wetted radius determines the change regulation of the flow rate. Matla has a powerful graphics processing function, using Matlab language to simulate the wetted radius of the square and triangle[20]. It can make the relationship between the wetted radius and other parameters more intuitive. Setting the maximum wetted radius as 1, the changing curve of the wetted radius for the square and triangle in polar coordinates and rectangular coordinates can be obtained through the Matlab software programming, as shown in Figure 2-26.

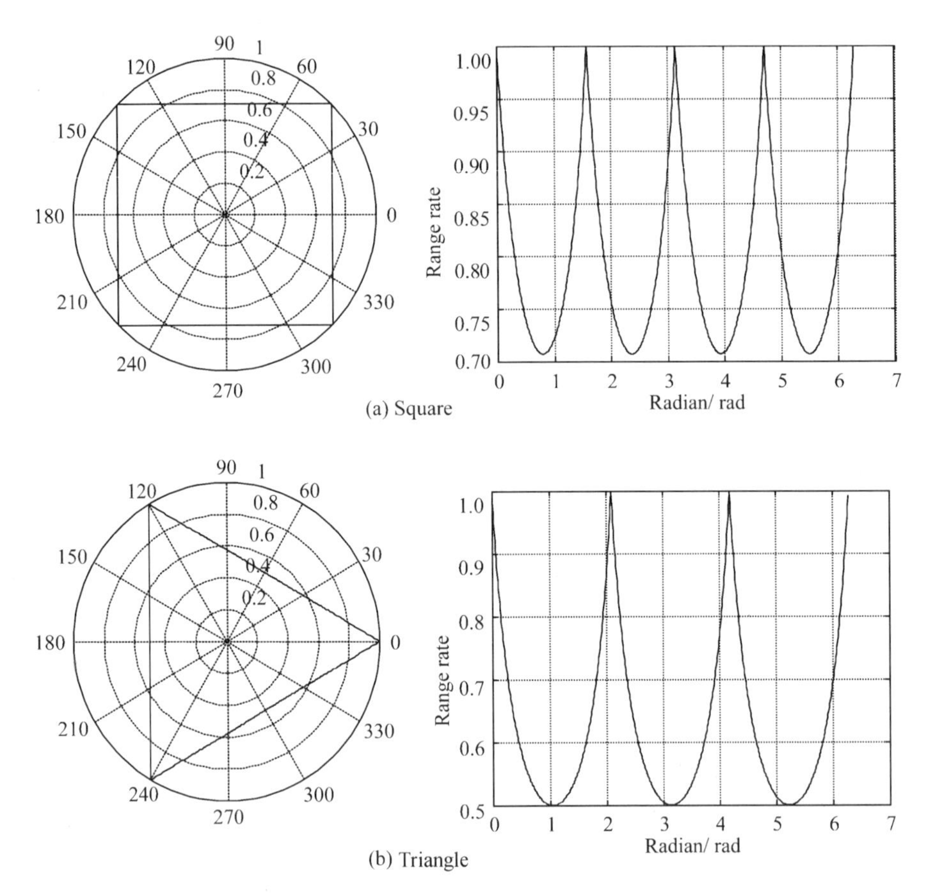

Figure 2-26　Changing curve of the theoretical wetted radius

The maximum wetted radius is set as 1, and the wetted radius is made equivalent. As seen in Figure 2-26, the maximum wetted radius outputs occur and vary through its different cycles, and the variational range of the triangle is relatively large. The correlation between the wetted radius and the flow rate determines the theoretical flow rate curve and corresponds to the wetted radius displayed in the square and triangle graphs of Figure 2-27.

(3) Relationship between the working pressure and the sectional area

Many factors impact the sprinkler's wetted radius[21]. Within the operating range, the effect of working pressure on the wetted radius is the most significant. Therefore, the wetted radius of the square and triangle can be changed by adjusting the inlet working pressure of the sprinkler.

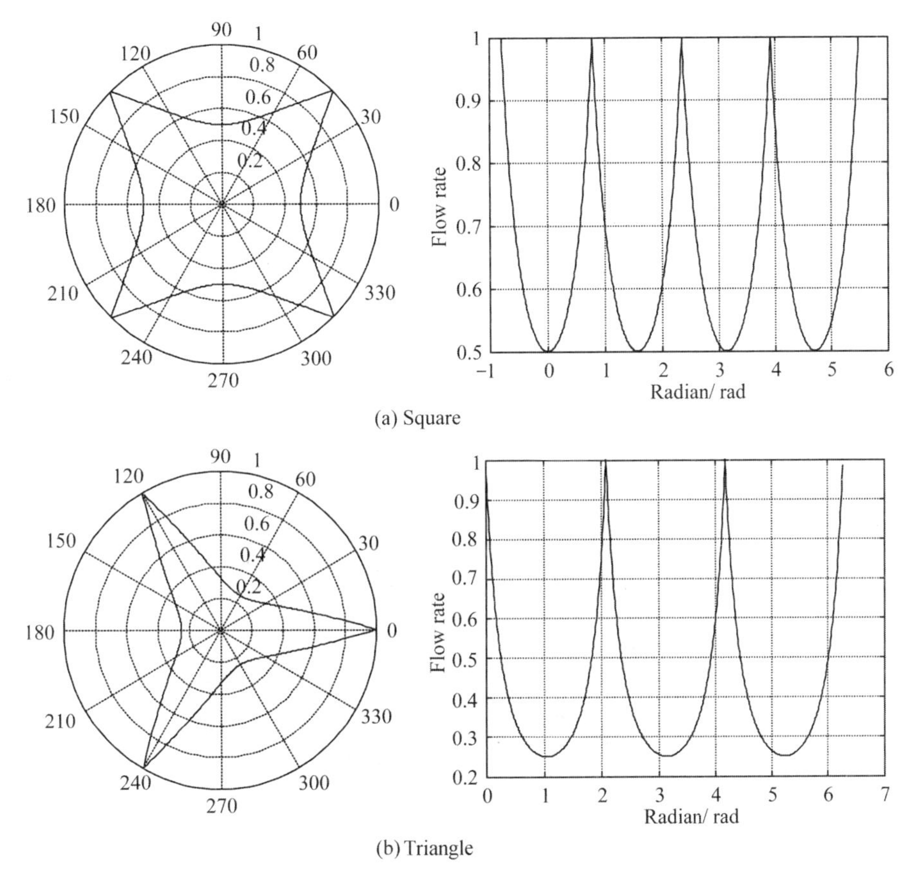

(a) Square

(b) Triangle

Figure 2-27 Theoretical changing curve of the flow rate

2.4.2 Types of variable-rate sprinklers

The fluidic sprinkler is a newly developed sprinkler, which has the advantages of a simple structure, excellent performance, and good application. Simultaneously, the traditional impact sprinkler is widely used in the market. Therefore, based on previously conducted research[22], the fluidic sprinkler and the impact sprinkler were selected to be used for the structural design and optimization of variable-rate spraying based on the pressure and flow rate regulating the device. A pressure and flow rate regulating device is added at the inlet of the sprinkler to change the pressure in order to optimize variable-rate spraying.

1. Variable-rate fluidic sprinkler

Variable-rate spraying for fluidic sprinklers is methodically conducted by installing a pressure and flow rate regulation device at the swivel of the fluidic sprinkler. The sectional area of the inlet changes during the relative motion of the dynamic and static pieces, altering the inlet pressure of the sprinkler, and altering the spray wetted radius[23]. The assembly drawing and product prototypes of a BPXH type variable-rate fluidic sprinkler are shown in Figure 2-28 and Figure 2-29 respectively.

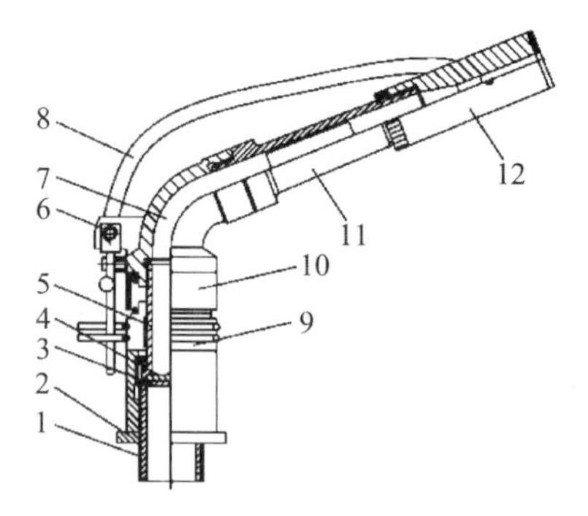

1—Connecting sleeve; 2—Swivel connecting seat; 3—Static piece; 4—Moving plate; 5—Hollow shaft; 6—Reversing mechanism; 7—Sprinkler body; 8—Reverse plastic tube; 9—Swivel; 10—Swivel seal mechanism; 11—Tube; 12—Fluidic element

Figure 2-28 Assembly drawing of the variable-rate fluidic sprinkler

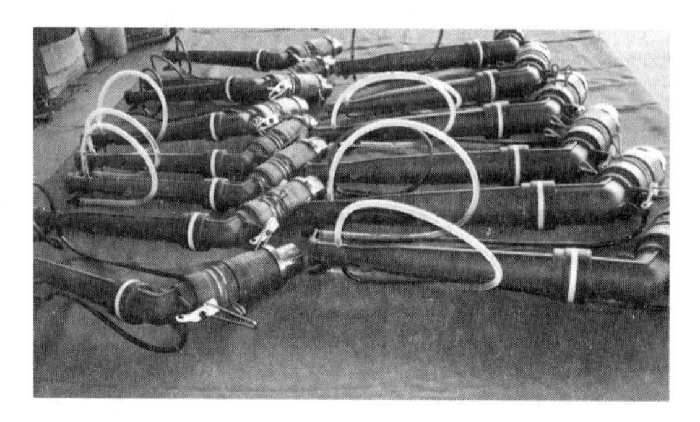

Figure 2-29 Prototypes of variable-rate fluidic sprinklers

2. Variable-rate impact sprinkler with non-circular nozzles

The impact sprinkler is widely used. Therefore, non-circular nozzles and automatically adjustable nozzles are used to match the pressure regulating device at the inlet. When the pressure regulating device determines the variability, the spray uniformity is improved, creating an improved hydraulic performance.

Designing non-circular nozzles, rather than relying on traditional, circular nozzles of the impact sprinkler aims to solve the problem of uneven dispersion of variable-rate spraying under the condition of a short wetted radius and low pressure. It leads to the improvement of water distribution and the spray uniformity of a single sprinkler.

The nozzle is an important part of a sprinkler, which directly affects sprinkler irrigation quality and hydraulic performance. It will not only maximize the flow pressure energy into kinetic energy but also maintain the flow through the current rectifier and have a low degree of turbulence, at the least not producing a large amount of transverse flow. Therefore, the main purposes of the design of non-circular nozzles are to determine the appropriate nozzle form and the optimal nozzle size.

Non-circular nozzles are new types of nozzles used in recent years both at home and abroad. It has the advantage of improving the jet atomization and water distribution of a single sprinkler. The concept of the non-circular nozzle was first proposed by the American Rain Bird Corp, which is called a "Control Droplet Size Nozzle"[24]. Data shows that by changing and making more reasonable geometry of a nozzle, the working pressure of a sprinkler is greatly reduced. Therefore, it has the possibility of increasing atomization and inevitably leading to a lower wetted radius.

According to the principle of the equal area between non-circular nozzles and circular nozzles, the dimensions of non-circular nozzles can be determined. Based on the PY_2 30 type sprinkler, four kinds of non-circular nozzle structures are designed, as shown in Figure 2-30. The star structure has three different dimensions. The circular section of the nozzle outlet guarantees the flow of the main jet and reduces the loss of the wetted radius as much as possible. Other parts like the non-circular design make the flow dispersed in the vicinity, and improve spray uniformity.

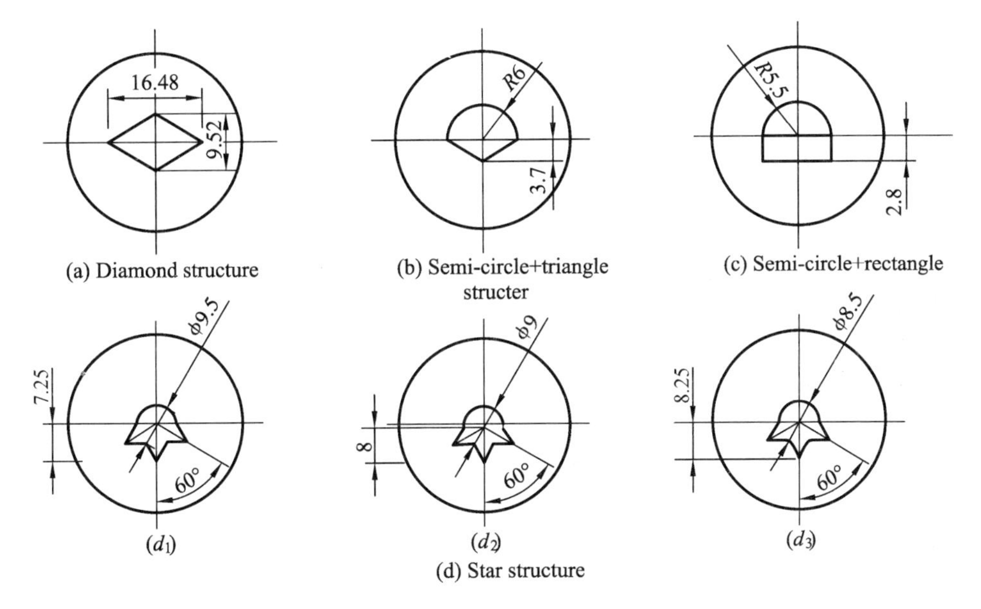

Figure 2-30 Structures and dimensions of non-circular nozzles

The wetted radius is related to the kinetic energy of the flow from the nozzle. The formula of kinetic energy are as follows:

$$W=\frac{1}{2}mv^2 \tag{2-29}$$

Because

$$m=\frac{\gamma q_P}{g} \tag{2-30}$$

$$v=\varphi\sqrt{2gH} \tag{2-31}$$

$$q_P=A\varphi\varepsilon\sqrt{2gH}=A\mu\sqrt{2gH} \tag{2-32}$$

So

$$W=\frac{\gamma q_P v^2}{2g}=\frac{\gamma A\mu\varphi^2\sqrt{2gH}}{2g}\cdot 2gH=\gamma\mu\varphi^2 AH\sqrt{2gH} \tag{2-33}$$

where W is the outflow kinetic energy in J, m is the outflow quality in kg, v is the outflow rate in m/s, γ is the severity of the water in N/m^3, q_P is the sprinkler flow rate in m^3/s, H is the water pressure head in m, μ is the flow rate coefficient, φ is the velocity coefficient, ε is the contraction coefficient, A is the nozzle area in m^2, and g is the gravity acceleration in m/s^2.

Formula (2-33) makes it evident that kinetic energy is proportional to the flow coefficient under the specific conditions of pressure and nozzle area. Therefore, in order to guarantee the size of the wetted radius of a sprinkler, it is necessary to select a non-circular nozzle with a larger flow rate coefficient.

3. Variable-rate impact sprinkler with the automatically adjustable nozzle

The variable-rate impact sprinkler with the automatically adjustable nozzle is a new kind of structure that affects an impact sprinkler's ability to calculate variable-rate spraying, as shown in Figure 2-31. It is composed of an inlet pressure regulating device, the automatic regulation mechanism of the outlet pressure, and impact sprinkler. The working principle is that when the sprinkler rotates, the inlet pressure regulating device installed in the swivel of the sprinkler changes the wetted radius of the sprinkler. When the water flows through the outlet automatic regulation mechanism, the water baffle plate automatically adjusts the outlet area of the nozzle and improves the dispersion of the water current in the conditions where there is low pressure and a short wetted radius, thereby improving the water distribution of the spray. The prototype of a variable-rate impact sprinkler with the automatically adjustable nozzle (BPY type) is shown in Figure 2-32.

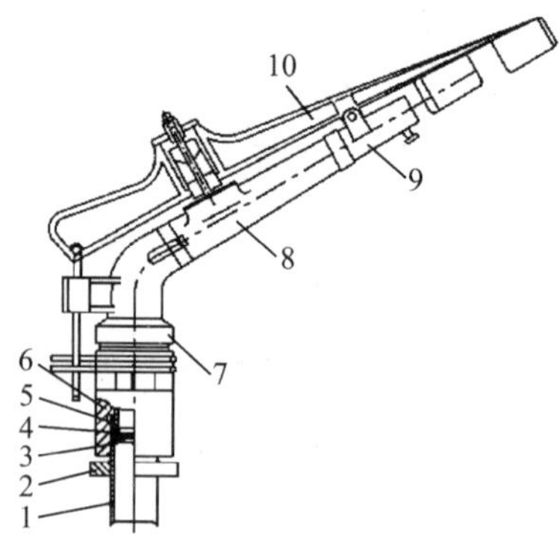

1—Sleeve; 2—Parallel; 3—Static piece; 4—Moving plate; 5—Hollow shaft; 6—Swivel; 7—Rotary sealing mechanism; 8—Tube; 9—Outlet pressure automatic regulating mechanism; 10—Drive arm mechanism

Figure 2-31 Structure of the variable-rate impact sprinkler with the automatically adjustable nozzle

Figure 2-32 Prototype of a variable-rate impact sprinkler with an automatically adjusted nozzle

Figure 2-33 is the schematic diagram of the automatic regulating mechanism of the outlet pressures. The automatic regulating mechanism of the outlet pressure is composed of a hollow screw, spring, water baffle plate, and outlet channel, and is connected to the outlet tube of the sprinkler. One side of the water baffle plate is fixed with a rotating shaft and is arranged in the flow channel; the initial position and the variation rule can be regulated by the spring arranged in the hollow screw. During the working time, the position of the water baffle plate changes with the flow pressure, and the outlet area of the nozzle changes. When the inflow pressure is large, the outlet area is large too, and vice versa. Thus, this changes the spraying uniformity automatically. Through the optimization design of the outlet flow channel, spring preload, and elastic modulus, the outlet area can be changed according to the required rule.

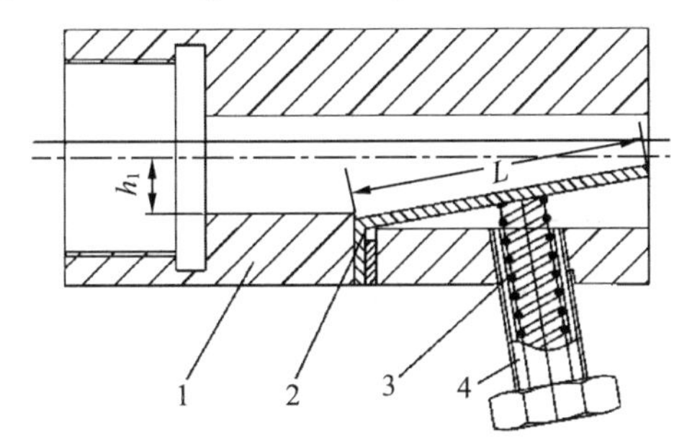

1—Outlet channel; 2—Water baffle plate; 3—Spring; 4—Hollow bolt

Figure 2-33 Schematic diagram of an automatic regulating mechanism of outlet pressure

2.4.3 Design method of the variable-rate sprinkler

There are many designs of pressure regulating devices for variable-rate sprinklers, which are widely used in irrigation systems. For example, Liu Xiaoli [25] tested the effect of pressure regulating devices on the hydraulic performance in drip irrigation systems. Tian Jinxia [26] studied the influence of pressure regulating devices on the outlet preset pressure in micro-irrigation systems. However, the research on the pressure regulating device mainly depends on the influence of the parameters on the performancen [27, 28]. These studies have not formed a set of design methods for the serialization and standardization of pressure regulating devices. Therefore, it is very important to design a set of practical methods to create a hydraulic pressure regulation device for variable-rate sprinklers. As exhibited through multiple designs and experiments, this chapter summarizes the design method of the arbitrary shape of the pressure regulating device. The design steps and roadmap are shown in Figure 2-34.

(1) Determining the type of sprinkler

Because the pressure regulating device is installed at the inlet of the sprinkler, the inlet size is very important. The general sprinkler inlet diameter is determined depending on the model of the sprinkler. Take a PXH30 sprinkler for example, which has a diameter of 30mm due to its specific design.

(2) Determining the excircle size of the movable plate and stable plate at the inlet of the sprinkler

Because a movable plate corresponds with a hollow shaft, the outer dimension of a movable plate should be determined by the size of the hollow shaft, which is set D_d. In the same way, the stable plate is fixed on the lower swivel; the connection size is set as D_j. Because the stable plate is connected with the swivel thread, the excircle of the stable plate is a thread.

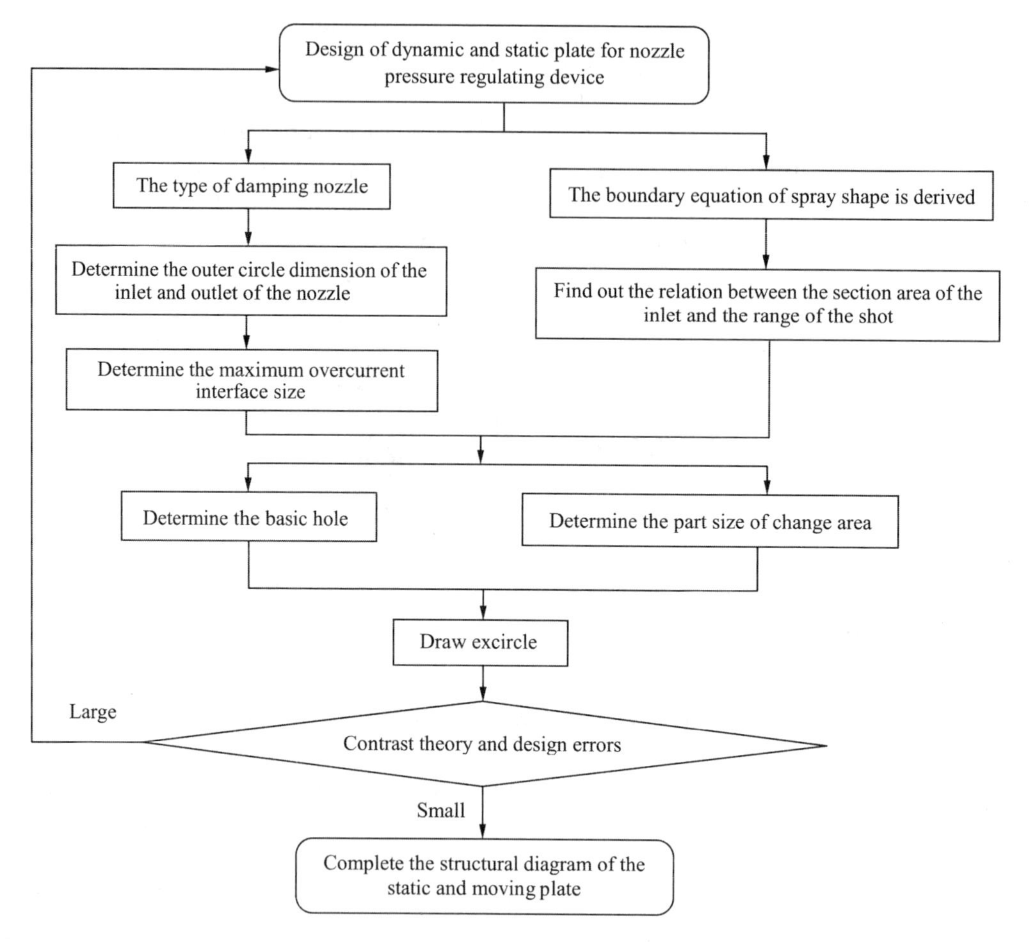

Figure 2-34　Flowchart of the design procedure

(3) Determining the maximum section size

Changing the location of a sprinkler will affect its wetted radius. To measure the change of the wetted radius under a different inlet sectional area, it is necessary to discover the change rule of the sectional area and wetted radius. The cut-off valves with different cross-flow areas need to be installed at the inlet, and the wetted radius should be measured three times to find an accurate average. When the inlet area of a section of a sprinkler is larger than that of a nozzle, the change of the wetted radius of the sprinkler is relatively slow. Therefore, in order to adjust the wetted radius continuously during the process of variable-rate spraying, the maximum inlet area of a section of the sprinkler needs to be set to the outlet area of the nozzle, which is $S_{max}=\pi\left(\frac{d}{2}\right)^2$.

(4) Writing the boundary equation of spray utilizing spray shape

Based on the shape of the spray area, the boundary equation of spray is established. Take a square and triangle for example, the change equation of the wetted radius with an arbitrary shape is set as $R=f(\alpha)$.

(5) Calculating the relationship between the inlet area and the wetted radius

The relationship between the wetted radius and the working pressure of each type of sprinkler is gathered through testing the wetted radius. In order to discover the continuity of the range adjustment in variable-rate spraying, the pressure range causing the most significant change is selected. In this range, the correspondence between the working pressure and wetted radius is fitted. The exponential fitting function is as follows:

$$y=ax^{b} \tag{2-34}$$

Using the change of the rotation angle, the inlet sectional area can be obtained.

(6) Determining the diameter of the basic hole on the movable plate and stable plate

According to the theoretical change rate of the area, when the rotation angle is zero and is set as the minimum sectional area of the nozzle, and the flow sectional area under this condition is set as the basic hole area. The area is $S_0=\mu_0 S_{\max}$. The radius of the basic hole is $r_0=\sqrt{S_0/\pi}$. Figure 2-35 is a schematic diagram of a movable plate and a stable plate.

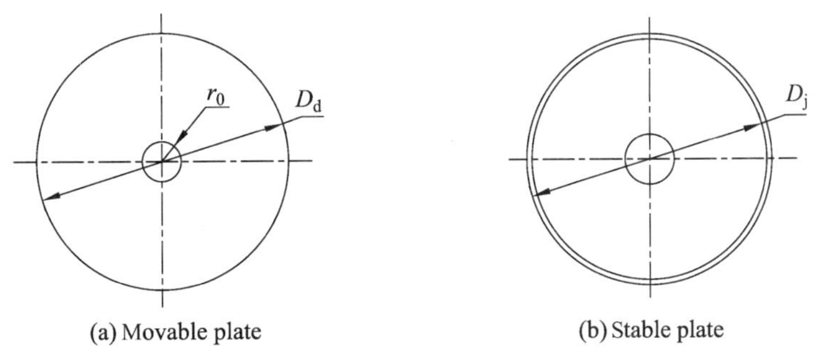

(a) Movable plate (b) Stable plate

Figure 2-35 Schematic diagram of movable and stable plates

(7) Calculating the size of the changing area

In calculating the change of an inlet area, the area change of the circumcircle of the basic hole is represented by S, where $S=S_i-S_0$, $i=1,2,3,\cdots$. Setting the radius of the circumcircle of the basic hole is $r_i=\sqrt{S/\pi}$. The distance between the center of the circumcircle and the center of the movable plate is set as $r_i'=r_i+r_0$. For example, position r_0 and position r_i of the movable plate of the triangular variable-rate spraying are shown in Figure 2-36 and Figure 2-37.

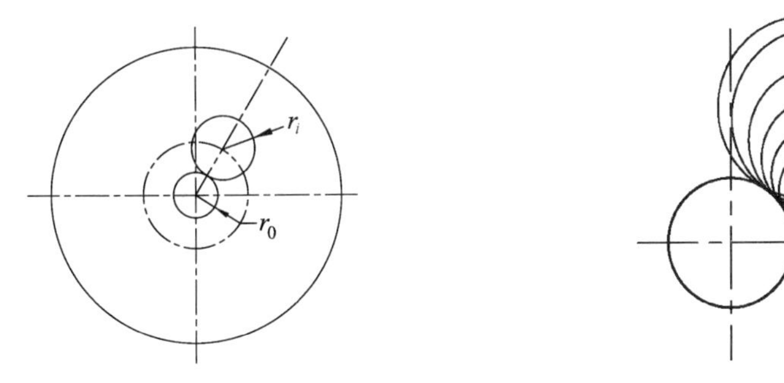

Figure 2-36 Relative position of the circumcircle **Figure 2-37 Sketch line of the circumcircle**

(8) Drawing the circumcircle using dimension

The circumcircle is drawn based on the distance between the circumcircle's center and the center of the moveable plate. A change rate within a range of 360 degrees is drawn by connecting each ray with the intersection of the circumcircle and smoothing the connection points. The structure of the stable plate is a circular hole in the middle. The circular hole is identical to the basic hole of the movable plate. Also, the relative motion of the movable plate and stable plate changes the inlet area. Therefore, the structure of the stable plate is obtained by extending the fan with a certain angle from the basic hole, as shown in Figure 2-38.

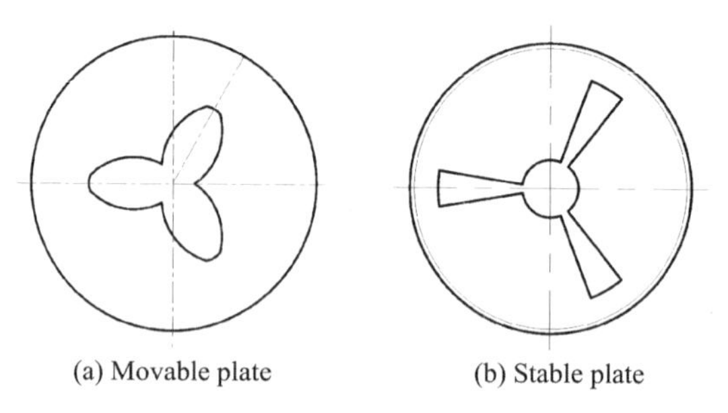

(a) Movable plate (b) Stable plate

Figure 2-38 Structural diagram of movable and stable plates

2.4.4 Variable-rate spraying overlap technology and its application

1. Overlap layout design

During engineering applications, the main factors affecting the layout of the pipeline system are topographic conditions, the shape of the plot, tillage and planting direction, wind direction and wind speed, and the location of the water source[29]. The shape of the spray area of sprinkler is also an important aspect of the sprinkler overlap arrangement. The sprinkler overlap principle is to ensure that spraying is done without a blank place, and has a high degree of uniformity. The determination of sprinkler spacing is a key step in the planning and design of a sprinkler irrigation system. Whether the sprinkler spacing is reasonable is not only related to the size of the investment, but also to the quality of irrigation crops, which affects the yield of the crop[30].

Based on the characteristics of the square spray area, and the optimal overlap principle of the spray, the layout can be arranged as shown in Figure 2-39.

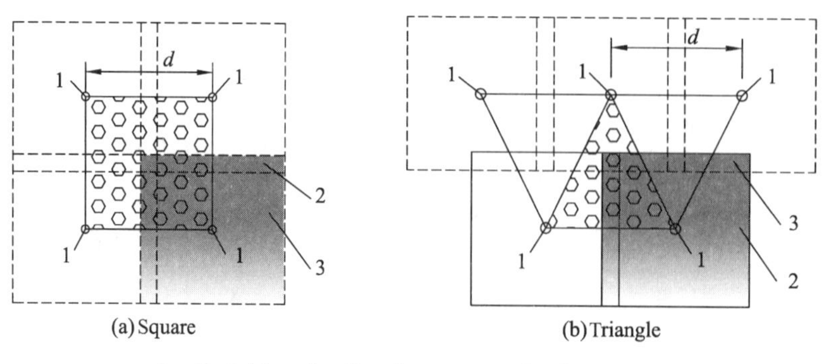

(a) Square (b) Triangle

1—Sprinkler; 2—Overlap square; 3—Spray area

Figure 2-39 Layout form

It can be seen in Figure 2-39a that two adjacent sprinklers have less overlap when the square spray sprinklers are arranged in a square layout. The part of the four nozzles is the numerical calculation area of the overlap uniformity of the spray. In the same way, it can be seen in Figure 2-39b that when the layout form is a triangle, the minimum wetted radius of the sprinkler overlaps with the maximum wetted radius of the other sprinkler and the area is small. The triangle area between the three adjacent sprinklers is the numerical calculation area of the overlap uniformity of the spray.

2. Calculation method of overlap uniformity

Sprinkler irrigation uniformity is the degree of water distribution in the irrigated area. It depends on the sprinkler spacing, the number of sprinklers running at the same time, and the water distribution of the single sprinkler. It is one of the key indicators of the quality of irrigation. When the water distribution is not uniform, it not only causes a varied crop growth and reduced crop yield but puddles and currents, leading to the waste of water resources and soil nutrient loss. On the other hand, the high quality of uniformity will result in a waste of energy and an increased cost of the system. A good operation scheme needs to ensure that the overlap water application rate is less than the water application rate that is best for the soil and the proper sprinkler uniformity.

When the water application rates of sprinklers are measured by grid arrangement, setting the point of measurement will be a $u \times v$ matrix: $\boldsymbol{E}=[e_{ij}]_{u \times v}$, where u is the number of rows of a matrix, v is the number of columns of a matrix, and e_{ij} is the element in matrix $\boldsymbol{E}$. Because of the square spray, then $u=v=n.i$ is the row i and j is column j. $\boldsymbol{E}=\boldsymbol{A}'+\boldsymbol{B}'+\boldsymbol{C}'+\boldsymbol{D}'$ In the formula, $a'_{ij}, b'_{ij}, c'_{ij}, d'_{ij}$ are the elements of $\boldsymbol{A}'$, $\boldsymbol{B}'$, $\boldsymbol{C}'$, $\boldsymbol{D}'$, respectively. Therefore, $e_{ij}=a'_{ij}+b'_{ij}+c'_{ij}+d'_{ij}$ $(1 \leqslant i \leqslant u, 1 \leqslant j \leqslant v)$. The adopted formula of the uniformity as follows[31]:

$$C_u=\left(1-\frac{\sum_{i=1}^{n}\sum_{j=1}^{n}\left|e_{ij}-\frac{1}{n \cdot n}\sum_{i=1}^{n}\sum_{j=1}^{n}e_{ij}\right|}{\sum_{i=1}^{n}\sum_{j=1}^{n}e_{ij}}\right)\times 100\% \tag{2-35}$$

Through Formula (2-35), the water application rate of the sprinkler irrigation at each point is measured by the collector arranged in the grid type substituted into e_{ij} of the formula, and the overlap homogeneity value is obtained.

2.5 Reaction rotating sprinkler

2.5.1 Structure of the reaction rotating sprinkler

There are many similarities between the structure of a reaction rotating sprinkler and other types of rotating nozzles, such as the rotary seal mechanism and channel. The driving part is the only difference and has various structures in different rotating sprinklers. However, the driving part of a reaction rotating sprinkler has only two types: the outlet of an inclined orifice (sloping, partial) and water baffle steering. The typical prototype of the reaction rotating sprinkler is shown in Figure 2-40.

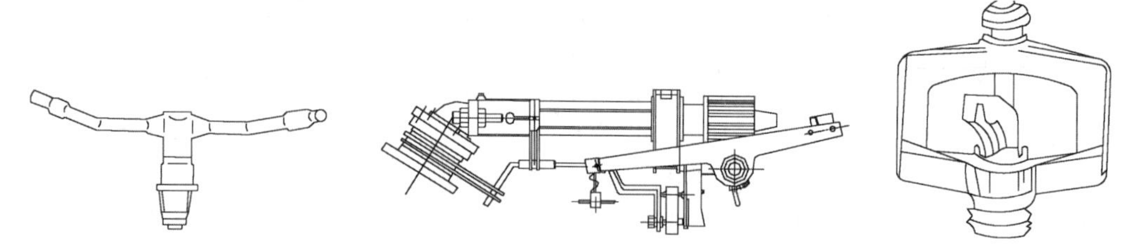

(a) Bend swing-arm reaction rotating sprinkler (b) Vertical impact sprinkler (c) Swing-arm reaction rotating sprinkler

Figure 2-40 Several typical reactions rotating sprinklers

For a small-sized sprinkler, it is the same as a fixed water sprinkler due to the similarity in an application. However, the reaction rotating sprinkler has a larger control area and low water application rate. The contradiction of the wetted radius is not very prominent. So, the inclined orifice discharge method is adopted to obtain the continuously applied reaction torque in order to make the sprinkler move continuously in full circular motion; at this point, the driving part of the sprinkler is very simple. For large and medium-sized sprinklers, the wetted radius is the main factor. The method of gap water baffle steering is used, and the reaction torque is applied in the gap to make the sprinkler rotate step by step. The process is a little more complicated, but compared to an impact sprinkler, it is much simpler. Therefore, a reaction rotating sprinkler is not inferior to an impact sprinkler from the view of the structure but is much simpler when compared with impeller sprinklers.

2.5.2 Water distribution of the reaction rotating sprinkler

In 1995, the United States Nelson Irrigation Company[32] developed an R3000 rotating sprinkler with an epoch-making significance. The key part of the sprinkler is a spinning disc with multiple channels, as shown in Figure 2-41. The working process of this sprinkler is first, the water flows out from the top of the nozzle, impacting the spinning disc to produce a rotational driving force, thus driving the spinning disc rotation. Playán[33] tested and compared the hydraulic performance of a rotating sprinkler and a refraction sprinkler and found that the water distribution of the rotating sprinkler was much better than that of the refraction sprinkler. The uniformity coefficient was more than 90%, which was an increase of about 16% compared to the reflective sprinkler. Also, when the installation height, working pressure, and nozzle diameter were the same, the wetted radius of the rotating sprinkler was larger than that of the refraction sprinkler.

Figure 2-41 Nelson R3000 spinning disc

At present, multi-channel technology has been applied by many companies, such as Nelson Company's A3000, B3000, and S3000 rotating sprinklers used in central pivot irrigators. Toro's PRN type rotating sprinkler. Rain Bird Company's RN type rotating sprinkler. Hunter's MP type rotating lifting sprinkler, etc. as shown in Figure 2-42, Figure 2-43, and Figure 2-44. However, the multi-channel structure is complex, difficult to manufacture, and has a high production cost.

Figure 2-42 PRN type sprinkler

Figure 2-43 RN type sprinkler

Figure 2-44 MP type sprinkler

The R33 rotating sprinkler produced by the Nelson Company in the United States is a low pressure rotating sprinkler with excellent performance. The sprinkler recognizes the overlap of water quantity in different wetted radius with the use of an intermittent water dispersion device and a damping control device. The non-circular nozzles and the intermittent water dispersion device can be used to improve the sprinkler irrigation quality at low pressure, shown in Figure 2-45. The working pressure of the sprinkler is about 50 kPa lower than that of the sprinkler with the same type.

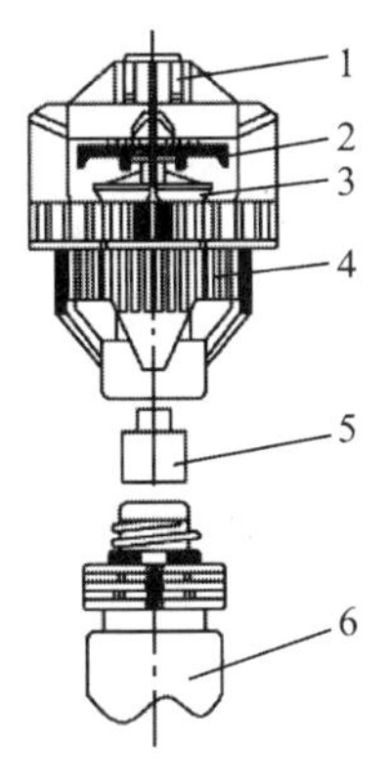

1—Damping structure; 2—Intermittent water dispersion device; 3—Spray nozzle; 4—Body; 5—Nozzle; 6—Connector

Figure 2-45 Schematic diagram of R33 sprinkler

2.6 Sprinkler test and test method

Sprinklers are the key equipment for the implementation of sprinkler irrigation operations. The performance of a sprinkler directly affects the sprinkler system's reliability and irrigation quality. This section describes the rotating sprinkler test, including the relevant test equipment, experimental methods, and test conditions.

2.6.1 Sprinkler test type and test equipment

1. Sprinkler test items[34]

Sprinkler experiments test the external characteristics, the internal characteristics, and the strength of a certain type of sprinkler. Test conditions can be conducted indoors and/or trialed in outdoor fields.

(1) External characteristics test

External characteristics examined usually include the testing of the operation of the sprinkler, the reliability of its rotation, and the uniformity of that rotation. External characteristics tests also include an examination of hydraulic performance and measuring the flow rate, wetted radius, spray height, water distribution, droplet diameter, droplet impact, and other performance parameters of the sprinkler.

(2) Internal characteristics test

The internal characteristics test mainly involves the evaluation of the rationality of various hydraulic structures of the sprinkler, such as verifying the rationality of the shrinkage angle of the sprinkler channel, the current regulator, the second nozzle, the sprinkler elevation angle, and so on. After the design of a rotating sprinkler based on the principle of the fluidic element, it is necessary to conduct the internal characteristic test to obtain the pressure distribution, velocity distribution, and jet switching condition of the water flow inside the sprinkler, as well verify the rationality of the design and find methods of improving the hydraulic design. The internal characteristic tests of sprinklers are generally used for new product developments.

(3) Strength test

The strength test of an agricultural rotating sprinkler is mainly carried out for evaluating durability and abradability. The strength of the developing production of unreleased sprinkler design, as well as mass-produced sprinklers that newly change their material, need to be tested.

2. The test equipment and methods of the sprinkler

The normal temperature of the water is generally sought during the performance test. The characteristics of normal temperature water should be consistent with the provisions of Table 2-11.

Table 2-11 Water standards for sprinkler test

Characteristics	Unit	The maximum value
Temperature	℃	40
Kinematic viscosity	m^2/s	1.75×10^{-6}
Mass density	kg/m^3	1 050
The content of free solids not absorbed water	kg/m^3	2.5
The solid content of water dissolved	kg/m^3	50

Sprinkler tests commonly use flow meter, pressure gauge, anemometer, hygrograph, thermograph, small weather station, raindrops spectrometer, collector, and many other pieces of equipment. The use of various types of measuring instruments should be accurate and capable of regular correction. The instrumental accuracy of a pressure gauge measurement is ±2% and the accuracy of the flow meter is ±1%. The accuracy of a timer used by rotation period and rotation uniformity is 0.1 s or more.

(1) Measuring the pressure

The measuring point of the pressure test is generally specified in front of a sprinkler's inlet at 200 mm. The pressure change in a test is usually not greater than

4%. There are different instruments of pressure measurement. One example is a U-shaped mercury differential pressure meter, which has no specific requirements for calibration. It can measure changes in positive and negative pressure conversions. There are shortcomings due to the limit of the instrument's height, which makes the general measurement range low. Another type of instrument is a spring pressure gauge. Spring pressure gauges use field display data, which are easily installed and easy to use. The third measuring instrument is a pressure transmitter. During a sprinkler test, it is necessary to pay attention to the working pressure range and the medium temperature range of the measured sprinkler when selecting a pressure transmitter. Going over the range will damage it. Transmitters that have gone unused for a long period need to be checked by the pressure calibrator.

(2) Measuring the flow rate

The flow rate measurement of a sprinkler test generally uses a turbine flowmeter, electromagnetic flowmeter, ultrasonic flowmeter, and other equipment. Small flow rate sprinklers, such as a micro-spray sprinkler, can directly collect water and the flow rate can be measured by utilizing the volume method and the weighing method. The pressure drop of a turbo flowmeter in sprinkler tests should be noted. The thinner the pipeline, the greater the pressure drop. At this time, the pressure gauge is generally installed after the flowmeter. Also, the process of installing a flow meter requires a straight pipe; the length must be ascertained before and after the installation. Compared to a turbine flowmeter, the electromagnetic flowmeter has no obstruction in its pipeline, no pressure loss, and it is easy to install a straight pipe. As with an electromagnetic flowmeter, an ultrasonic flow in the pipeline has no possible obstruction and no pressure loss. The outstanding advantage is that it is barely affected by the various parameters of the media (temperature, pressure, viscosity, density, etc.). During a unit test of outdoor sprinkler irrigation, the flow rate inside the pipeline can be easily measured.

(3) Measuring the wind speed

Instruments measuring air flow rate can be a wind cup, propeller, hotlines, and so on. In an outdoor sprinkler irrigation experiment, the wind speed greatly impacts the water distribution of a sprinkler, effecting the drift of the small water droplets. In general, when winds are categorized at or above the third level, the outdoor water distribution test is ineffective and skewed. The wind speed and direction should be measured at a set interval of time, ensuring that the interval length is no more than 15 minutes. The time will be specific to the testing time of a single sprinkler or an irrigator. When using an anemometer, the testing height is ninety percent of the maximum trajectory height of a sprinkler no less than 2 m and no more than 50 m from the edge of the test site. The wind speed and direction data should be described in the test report and the data should be measured and collated for several times.

(4) Small weather station

Small weather stations are used for monitoring weather features on-site such as wind speed, wind direction, rainfall, air temperature, air humidity, light intensity, soil temperature, soil moisture, evaporation, atmospheric pressure, and so on. The configuration of meteorological observation elements can be flexibly configured according to the actual situation of the project and can be connected to a computer through the professional data acquisition communication line, which transmits the data to the meteorological computer meteorological database. Small weather stations have an important role in detecting environmental changes in outdoor sprinkler irrigation experiments, and carefully studying transpiration, evaporation, and soil infiltration of the sprinkler irrigation water.

(5) Collector

Spray water volume is gathered by the collector in the sprinkler irrigation test, and then the water distribution of the sprinkler or the irrigator is analyzed. The upper part of the collector should be cylindrical, the shape and size should be uniform, the opening edge should be sharp, and there should not be any gap. According to the provisions of the specifications, the height of the water-receiving cylindrical part is at least 1/3 of the height of the collector. The height of a collector is twice as much as the average height of water collected during the test and is not less than 150 mm. To prevent splashing both internally and externally, the diameter of a collector should be 1/2 to 1 time of its height, but not less than 85 mm.

Collectors should be placed vertically; the height difference between two adjacent collectors should not be more than 20 mm. The allowable inclination between the opening of the collector and the horizontal is $\pm 5°$.

Figure 2-46 is a structure diagram of a skid-type automatic metering collector. This kind of collector is composed of a shell, base, upper tipping bucket, metering tipping bucket, count funnel, and water receiver.

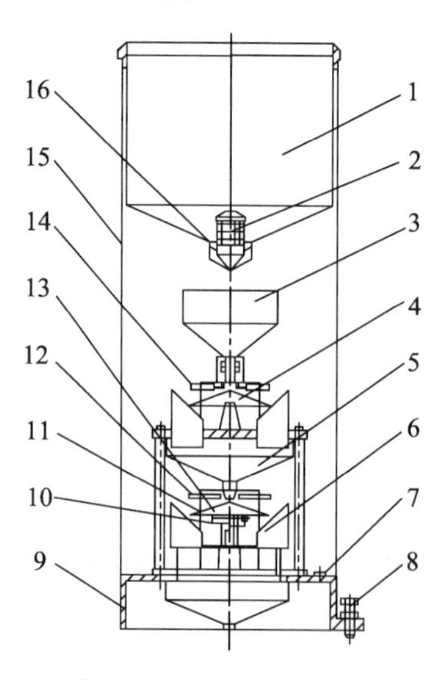

1—Water gauge receiver; 2—Mesh enclosure; 3—Funnel; 4—Upper tipping bucket; 5—Collection funnel; 6—Count funnel; 7—Spirit bubble; 8—Adjusting nut; 9—Base plate; 10—Reed switch; 11—Binding post; 12—Metering tipping bucket; 13—Capacity adjusting screw; 14—Set screw; 15—Cylinder; 16—Removable nut

Figure 2-46 Schematic diagram of the automatic metering collector

Spray water is received by the water receiver, leaking into the upper tipping bucket, and then enters the metering tipping bucket. The metering tipping bucket has a filling capacity of water, due to changes in the center of gravity, to transfer the water into the count funnel. The metering tipping bucket flips once, indicating a certain amount of precipitation (depending on the minimum size determined by the manufacturer). A small piece of magnetic steel is installed on the count funnel. When the count funnel from a stable position to another stable position is flipped with a magnet swept by the reed switch, two normally opened reeds are attracted in the reed pipe. At the moment of contact, the external switch circuit forms a loop and is connected once, and the circuit voltage is reduced once to form a pulse signal, which is sent to the acquisition system to obtain rainfall information.

The form of collector can be varied, but regardless of its configuration, it needs to ensure its measurement accuracy. The two kinds of collectors used in the sprinkler irrigation laboratory at Jiangsu University are shown in Figure 2-47. The left side of the figure is the automatic output signal collector, characterized by the use of a double-tipping bucket structure and electronic trigger switch. Rainfall data is in wireless transmission, so the collector can be placed anywhere. The minimum measurement of precipitation is 0.1 mm. The right side of the figure is the invention of the new collector. It is easy to read data through the conical measuring cup with a scale. It is convenient for mobile outdoor testing, made of transparent material, and the base support weight increases its windproof capabilities.

Figure 2-47 Two kinds of collectors

(6) Measuring the droplet diameter

There are many methods to measure droplet size, and each has its advantages and disadvantages. At present, the methods of measuring droplet diameters include the laser precipitation monitor, filter paper method, flour method, and high-speed photography method.

① Laser precipitation monitor. The laser precipitation monitor (LPM) manufactured by Thies Clima, a German company, can be used to measure the number, size, intensity, and exact speed of the high-speed moving objects. Its superior performance is particularly manifested in the measurement of tiny objects with a minimum diameter of 0.16 mm. For the measurement of rainfall, the amount, speed, particle size, intensity, and grade can be obtained by using an LPM. Thus, it was used in this research to measure droplet diameters of the aeration impact sprinkler and the original impact sprinkler at the end and in the middle of the wetted radius. The LPM system includes two subsystems: (ⅰ) an imaging system composed of a photodiode detector, laser transmitter, storage circuit, and others, as shown in Figure 2-48. In this particular example, the laser size is 785 nm, the test area is 46 cm^2 (23.0 cm × 2.0 cm), the measurement range of the particles' diameters are between 0.16~8.00 mm, the velocity range of the particles is 0.2~20.0 m/s and the measurement range of the rainfall intensities is 0.005~250.000 mm/h; (ⅱ) an analysis

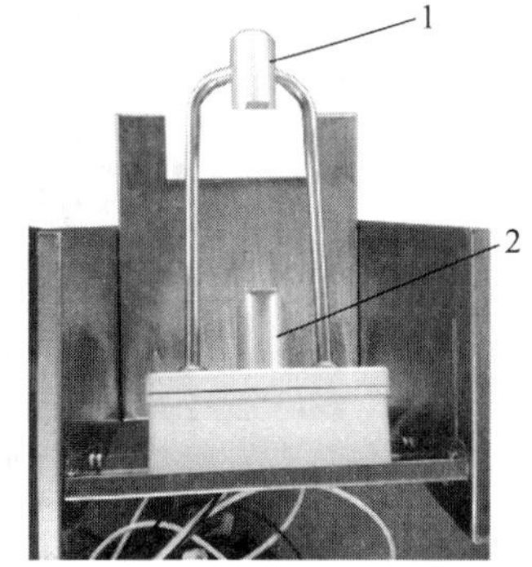

1—Photodiode detector;
2—Laser source

Figure 2-48 Structure of LPM

display system composed of LNM View software which was used to display data generated by the LPM. It can accumulate measured droplet volumes and calculate the rainfall intensity. The droplet spectra of LPM can be drawn, containing the drop size range, the droplet speed range, and the corresponding particle numbers. The data collected per minute could be output to an EXCEL file.

② Filter paper method. The filter paper method is a method to calculate the droplet diameter by measuring the splash diameter formed on the filter paper coated with toner, and then calculating the droplet diameter according to the formula of pre-filtration. There is also dissolved carbon tetrachloride in the methyl blue suspension sprayed on the filter paper to get droplet stains. Toner is usually mixed with red reagent and talcum powder with a 1 : 10 ratio. Filter paper is fixed on a water receiving box, which samples a few droplets during the experiment. When the droplets have left a stain on the filter paper, the drops can be dried. Then, based on the pre-filter diameter and the diameter of the curve or formula, the diameter of the water droplets can be determined.

For qualitative filter paper with a uniform texture, the parabolic relationship between the stain and the droplet as follows.

$$d = aD^{b} \tag{2-36}$$

where d is the droplet diameter in mm, and D is the stain diameter in mm.

Coefficient a and index b vary with the filter paper, so each batch of filter paper needs a filter set before the test to determine the value a and value b. The filter paper method is simple and easy to use. However, due to the randomness of the received water droplets, the measurement accuracy is determined by the caution taken by each tester, and the manual workload is large. Figure 2-49 is the droplet calibration chart and photos of the measured droplets.

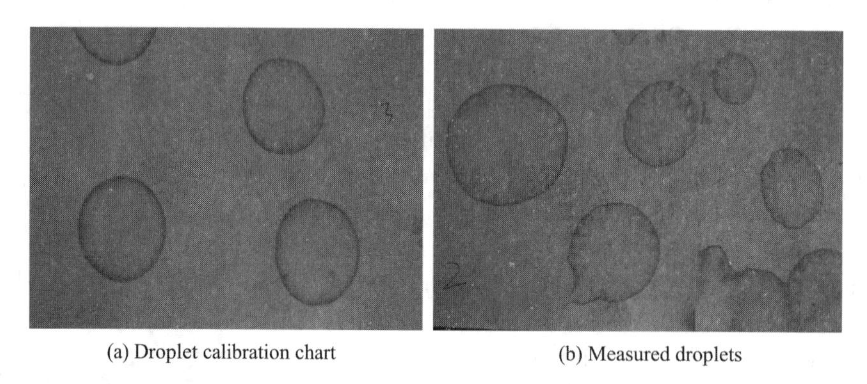

(a) Droplet calibration chart　　(b) Measured droplets

Figure 2-49　Droplet calibration chart and photos of the measured droplets

③ Flour method. The flour method encompasses sieving fresh white flour and putting it onto a dish with a diameter of 21 cm and depth of 2 cm, flat with a ruler. The ready-made flour dish must be tested within 2 hours. In the test, the flour dish is put in a sampling position. The flour dish is then taken back and placed in an oven after sampling, drying at 38 ℃ for 24 hours. In order to eliminate the possibility that droplets may be cut by the sharp edges of a dish, only a sample of 18.3 cm in diameter is taken from the center of the dish. The sample needs to be placed in a 50 mesh sieve, so the dough formed by the droplets is separated from the flour. Then 16 meshes are screened with 5～50 meshes and weighed respectively.

The ratio R_a of the droplet quality and the dry flour mass M_p is measured by releasing the droplets with known mass into the flour dish from different heights. Landing height is from 0.1 to 4 m, which has almost no impact on the R_a. For droplets with diameters of 2.19 to 5.32 mm, the R_a values are the same. The relation between R_a and M_p as follows:

$$R_a = 1.05 M_p^{0.082} \tag{2-37}$$

④ High-speed photography. The use of a high-speed camera functions to shoot photos in a very short period to complete the moving target quickly and multi-sampling, with a high-frequency recording. The recording speed ranges from 1 000 frames per second to 10 000 frames per second. A complete set of high-speed imaging system consists of optical imaging, photoelectric imaging, signal transmission, control, image storage, and processing of several parts. High-speed photography technology has advantages such as real-time target capture, fast image recording, real-time playback, visual clarity, and other prominent advantages. Using high-speed photography to obtain the pictures can determine the relationship between the droplet diameter and the drop angle of droplets, which can also capture the trajectory of droplets. Figure 2-50 is the photos of droplets measured by high-speed photography. From the photos, the droplet sizes can be measured and estimated by setting an accurate reference scale.

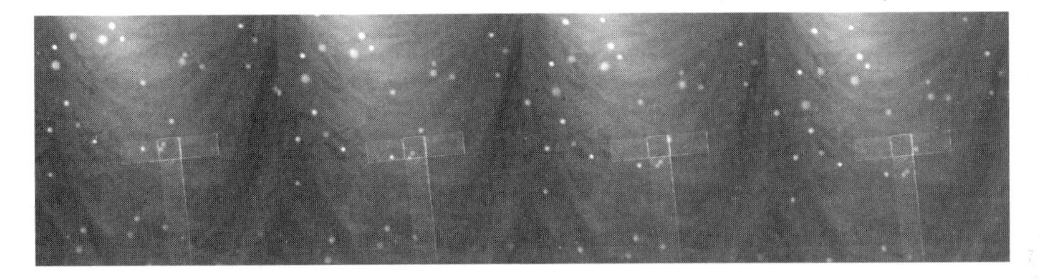

Figure 2-50 Photos of droplets measured by the high-speed photography

It can be seen from the above figure that the high-speed photography method is quick and easy, the droplet size in the picture needs to obtained with the image post-processing technology, so the method is worthy of further study.

2.6.2 Test methods of the sprinkler

1. Pressure test[35, 36]

In order to investigate the pressure resistance of a sprinkler during the test of sprinkler quality, it is usually divided into a warm water test and a hot water test. In an experiment, the sprinkler is installed on the test device, and the nozzle is blocked. At the beginning of the test, the residual air in the system is removed and then begins from 1/4 of the rated working pressure. The test pressure is gradually increased to the required maximum test pressure. The test pressure of the metal sprinkler at normal temperature water is 2 times the maximum rated working pressure, and the pressure is kept for 10 minutes. In the test, the sprinkler body and the tube (not including the rotating bearing) will not be damaged or leaked. The hot water test encompasses soaking a sprinkler in water with a temperature of 60 ℃. The test pressure is the maximum rated working pressure, which is kept for 1 hour for metal sprinklers and 24 hours for plastic sprinklers. The sprinkler and its parts should not be damaged or fall off, and the sprinkler body and screw joints should not leak.

2. Seal test of the sprinkle bearing

The test is carried out after 24 hours of operation under the maximum working pressure of the sprinkler. According to the normal condition, the sprinkler is installed on the water supply pipeline. Pressure should then be increased in increments of 100 kPa, from the minimum working pressure to the maximum; each increment should be maintained for one minute. The water that leaks from the sliding bearing should be collected throughout the entire test to check whether it was leakage within the specified limits. There are specific requirements regarding the tightness of a sprinkler. The leakage of a rotary bearing should not be greater than 0.005 m^3/h when a sprinkler's nominal flow rate is less than 0.25 m^3/h. In general, the leakage of rotary bearings should never exceed 2% of the sprinkler's nominal flow rate, ranging from 0.25 to 5 m^3/h. If the sprinkler's nominal flow rate ranges from 5 to 30 m^3/h, the leakage of the bearings should not exceed 1%. Finally, in a third scenario, the leakage of the rotary bearings should not be over 0.5% of a sprinkler's flow rate if it is greater from 0.25 to 5 m^3/h.

3. Seal test of the nozzle connection

During an experiment, a sprinkler's nozzle hole is blocked without using sealing materials. During the test, the air in the system is removed firstly, and then the minimum working pressure is increased to 1.6 times the maximum working pressure. The pressure is kept for 10 minutes at room temperature. During the whole test, the water leaked from the nozzle connection was collected to see if it was within the specified limits.

4. Rotation reliability test

Before the test, a sprinkler is soaked in water at the temperature 60 ℃ for 1 hour and then arranged in a vertical riser. The water pressure will gradually rise from zero, leading to the sprinkler's smooth rotation in one direction. This occurs over two minutes. The pressure will eventually increase to the maximum working pressure, running for 1 minute at this pressure. To deviate the sprinkler axis from the line, it is tilted 10 degrees, as shown in Figure 2-51, and the test is repeated. The purpose of such a test is to examine whether or not a sprinkler will operate normally in the range of its minimum and maximum working pressure in a tilted state.

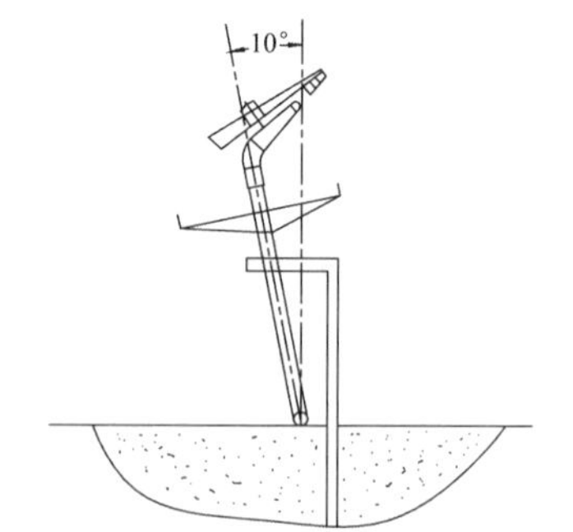

Figure 2-51 Tilt direction of the sprinkler rotating axis

5. Rotation uniformity test

During the experiment, the sprinkler is installed on a vertical riser, running a rated working pressure. The rotation time of four quadrants in a week is measured, repeated 5 times, and then the average rotation time of each quadrant and the maximum deviation with the mean value are calculated to see whether it is within the prescribed range.

6. Hydraulic performance test of the sprinkler

The hydraulic performance test is the most important method to test a sprinkler's quality, including the following steps:

① Measuring the sprinkler's flow rate. The above-mentioned flow rate measurement equipment is used to measure the flow rate of water sprayed from a sprinkler outlet. The sprinkler outlet includes main and secondary nozzles, non-circular nozzles, and a channel for a non-rotating sprinkler. For a sprinkler with a flow rate under 0.25 m^3/h, the change of the sprinkler flow rate should not be greater than $\pm 7\%$ at the specified test pressure. For a sprinkler with a flow rate greater than 0.25 m^3/h, the change of the flow rate at the specified test pressure should not be greater than $\pm 5\%$.

② Measuring the point water application rate. To measure average irrigation depth on an irrigation area per unit of time, the water volume at different positions is mainly collected by a collector, and the water depth can be obtained when the collected water volume is divided by the cross-section of the collector. Then, the water depth is divided by the test time to obtain the point water application rate. The uniformity of the water distribution of the sprinkler is reflected by the water application rate of each measuring point.

③ Measuring the trajectory height of a sprinkler. Depending on the height of irrigation crops, combined with the installation height of a sprinkler, the trajectory height needs to be considered. Especially when the large spray gun is applied to dust suppression, cooling, fire protection, and other sprayings, in which the elevation angle of the sprinkler has a greater impact on the trajectory height. The measurement of the trajectory height can be done by using a theodolite and setting photographic method of the reference object. When the sprinkler operates under the test pressure, the maximum rated working pressure, and the minimum rated working pressure, the vertical distance between the highest point of the spray water, and the horizontal line of the nozzle is measured respectively.

④ Measuring the wetted radius of a sprinkler. A sprinkler's wetted radius refers to the distance between a sprinkler and the prescribed point water application rate. In the continuous operation of a sprinkler, for a sprinkler with a flow rate greater than 0.075 m^3/h, the water application rate of this point is 0.25 mm/h; for a sprinkler with a flow rate equal to or less than 0.075 m^3/h, the water application rate of this point is 0.13 mm/h. Commutation sprinkler should be measured at any angle other than at its maximum limit angle. The sprinkler wetted radius measurement is carried out simultaneously with the water distribution characteristic test.

⑤ Measuring the end droplet diameter and the impact force. The impact force of a droplet affects on soil compaction; it is easy to scour and form a small path flow when the impact force is too large. The early growths of part crops cannot be impacted too forcefully. Generally, using the micro-pressure sensor to measure the rainfall pressure, the measured values are related to rainfall density. Also, the droplet impact force can be obtained by combining the droplet velocity and particle size measured by the laser precipitation monitor.

7. Endurance test of the sprinkler

The schematic diagram of the endurance test device is shown in Figure 2-52. The test was carried out under the maximum effective working pressure (the maximum rated working pressure). The provisions of the "rotating sprinkler GB/T 22999—2008" standard requires the total working time shall not be less than 2 000 hours. For a sprinkler with a commutator, the reliability assessment time of the reversing mechanism is not less than 1 000 hours. For fixed sprinklers, there are no specific provisions. The size of the main parts should be measured before the test, and the wear and corrosion of the parts should be checked after the test. During the experiment, the sprinkler continuously ran for 4 to 5 days in the test bench, and then was stopped for 1 to 2 days, because of the law of alternating (a sprinkler operates during a specified time). Someone needs to be dedicated to keeping records during the test. After the endurance test, the pressure test, sealing test, rotation uniformity test, sprinkler flow rate test, and water distribution characteristic test should be repeated to check whether they meet the requirements.

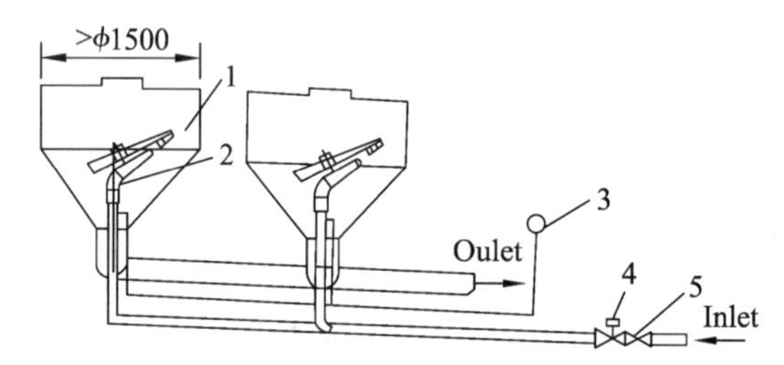

1—Water collecting tank; 2—Sprinkler; 3—Pressure gauge; 4—Pressure regulating valve; 5—Quick opening and closing valve

Figure 2-52 Schematic diagram of the endurance test device

8. Water distribution uniformity test

The water distribution uniformity reflects the performance of a single sprinkler which is the basis of the overlap sprinkler irrigation with multi sprinklers. The site of the collector placed during the test needs to be flat; the maximum allowable slope is 1%. There should be no obstacles to the free distribution of the spray water in the test field. In order to avoid the airflow above the field test, a field test can be set in a sealed and airtight room, and also can be located in outdoor open space away from trees or tall buildings. For a full circle sprinkler, the test time should be no less than 1 hour. For a reversing sprinkler with s fan spray, the shortest allotted test time as follows:

$$T_s = \frac{t\varphi}{360} \tag{2-38}$$

where T_s is the duration time of the fan spray in h, t is the duration time of the full circle spray, and φ is the actual fan angle of the reversing sprinkler (°).

The layout of collectors has two arrangements in a sprinkler water distribution test; one is a grid arrangement and the other is a radial arrangement, as shown in Figure 2-53.

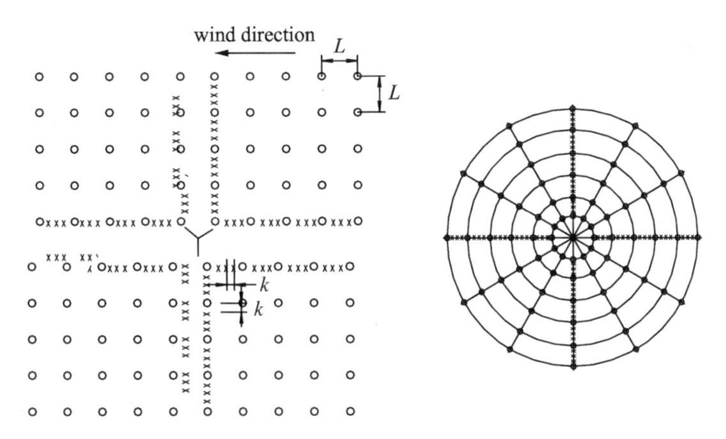

Figure 2-53 Square and radial arrangements of collectors

The area of the collector placing site should be larger than the spray coverage area of the sprinkler, and the appropriate safety margin should be left. Layout spacing varies depends on test items and a sprinkler's wetted radius. When the effective diameter of coverage is greater than 10 meters, the length of L in the figure is set as 2 meters. When the effective diameter of coverage is less than 10 meters, the length of L in the figure is set as 1 meter. When measuring the wetted radius of the sprinkler, the distance of k in the figure is set as 0.5 meters, and the allowable deviation of the distance between the collectors in any direction should not be greater than 50 mm. The grid layout method makes it easy to carry out the overlap sprinkler irrigation test examining multiple sprinklers. At this time, the layout position of the sprinkler is determined by the overlap mode of the sprinkler. The grid arrangement is the most appropriate choice when developing a non-circular sprinkler test. The radial arrangement is suitable for conditions where the wind is not a factor to measure the wetted radius of a sprinkler. The collector is arranged in a radial direction with the sprinkler as the endpoint. When the effective diameter of coverage is greater than 20 meters, the largest distance of the collector is set as 2 meters; when the effective diameter of coverage is less than 20 meters, the distance is set as 1 meter.

During the test, it is vital to take note of the starting and stopping positions of a sprinkler's rotation, as it will have an impact on the measurement results. To ensure that the collectors in each row receive an equal amount of rainfall, the starting and ending positions are set to a blank space between two rows of collectors. During the experiment, immediately following a pump stopping its operation and water leaving a sprinkler, its unstable stage should be noted. At this time, the water from the sprinkler should not fall into the collector. When adjusting the working pressure of the sprinkler or changing the working parameters, the spray water should be placed in a neutral position or the water will be blocked by the appliance. It will be released after the pressure is adjusted.

9. Slope test of the sprinkler

During the actual sprinkler irrigation project, the terrain with a certain slope influences the water distribution of the sprinkler and is not ideal for the ground. In order to prevent the formation of runoff on sloping land, which allows the water application rate to decrease, the study of the sloping land of the sprinkler application needs to carry out corresponding sprinkler tests. The test is to verify the theory between the sloping land and the sprinkler, looking for an important relationship that will determine the slope irrigation strategies and methods. For example, if the condition of the ground slope is 6.6 degrees, the sprinkler wetted radius compared to the ground wetted radius will have an 8% test error and 12.4% theory error. The hydraulic performance test of a single sprinkler on a sloping field is different from the ground on the layout of the collector; the method of artificial slope simulation is more direct and convenient. Placing the collector on stakes or supports and placing them according to the slope rate will lead to more accurate measurements. Placing the sprinkler in the middle of the slope, carrying out the spray irrigation test on upper and lower sides in two directions at the same time, or to minimize the height of the stakes, the sprinkler can be placed at the bottom of the slope to carry out uphill rainfall test, then placing at the top of the slope to carry out the downhill direction precipitation test. The requirements and regulations for the collector spacing, sprinkler test time, and wetted radius are the same as the ground level test. When the simulation slope is greater than 7 degrees, the angle between the sprinkler vertical pipe and the horizontal direction, as well as the elevation angle of the sprinkler, needs to be recorded. Then, the angle of the sprinkler vertical pipe needs to be changed to the appropriate position so the rotation uniformity test of the sprinkler can be conducted.

2.6.3 Test design and report

1. Test design

In order to determine the effect of a structural parameter on the performance of a sprinkler during its design, some separate experimental studies need to be conducted. Before testing, one must determine the design of the experiments scientifically and reasonably. The test design is not only the basis of scientific experiments but also the key to the success of scientific experiments. A good test design reflects the following aspects:

(1) A clear purpose

The test design is mainly based on the purpose of the experiment in understanding a phenomenon or a series of problems. For example, the purpose of a test could be to compare the effect of a component before and after the improvement, and then to compare the test results before and after the improvement without having to involve other test factors and analysis. If an experiment is to explore multiple test factors affecting the test indicators and their regularities, then we should consider the multi-factor test design and detailed analysis of its rules.

The simplest test design method should be used as much as possible to accomplish the test task with the least manpower, material resources, financial resources, time, and test numbers to achieve the purpose of the experiment.

(2) Control test factors and test conditions

The test design principle is used to improve test accuracy and reduce test error.

(3) Determination of test design methods

In a multi-factor experiment, an orthogonal experiment design is often adopted.

Through an orthogonal table, a representative method of partial combination treatment is selected, and a scientific test method is selected. The orthogonal experimental design is the application of a balanced analysis of test design principles, so the orthogonal test design and some flexible application methods can also be used for all combinations of multi-factor tests.

The orthogonal test design can be summarized as follows:

① The test numbers can be reduced, the efficiency can be improved, and the scientific research and test cycle can be shortened. For an experiment with multi-factors, some orthogonal experiments can be made, and some orthogonal experiments are arranged to select a better combination treatment. If necessary, it can be processed around a better combination, and continue to do a second or more batch of experiments to select the best combination treatment, to achieve excellence from accuracy.

② On the complex multi-factor test, the examination is more comprehensive; one can distinguish the main and secondary effects and trends between the experimental factors on the impact of the test indicators, which can seize the main contradiction. Also, the optimal combination of the optimal level in each factor of the test can be selected.

③ The method is simple: arrange the experiment according to a set of tabular orthogonal tables. The intuitive analysis is satisfactory as long as simple arithmetic operations are used for analysis.

2. Test error analysis[37]

The test error of sprinkler mainly includes two aspects. One is the error produced by data measurement, and the other is the error produced in the data processing. In the process, errors can include an insufficient understanding of measurement, or the contact method disturbing the original state. The static method solves the dynamic problem which can bring measurement errors and so on. Factors of the environment (temperature, atmospheric pressure, electromagnetic field, etc.) can also bring about measurement error. The error of data processing can be caused by an effective digit and approximate calculation.

The errors can be categorized as random or systematic. The method of0 reducing random errors mainly depends on improving test methods and improving measuring techniques. The processing system error requires strong pertinence, according to the actual processing. After studying and checking the factors that affect the measurement results, effective measures are taken to limit the system errors and reduce the system errors.

References

[1] Rural Water Conservancy Department of the Ministry of water resources, China irrigation and drainage development center. Handbook for use of water saving irrigation engineering[M]. Beijing: China Water Power Press, 2005.

[2] Xiang Q J, He P J, Lu H Q.Maximum inspiratory volume of the Water jet air ejector[J]. Journal of Agricultural Machinery, 2006,3: 145 - 148.

[3] Lu H Q. Theory and application of jet technology[M]. Wuhan: Wuhan University Press,2004.
[4] Liu J R, Zhou G P, Shi W D.Design and study of light and small jet self-suction spray pump[J]. Water Pump Technology, 2006,3: 1-4.
[5] Liu J P, Zhu X Y. Key technology of complete jet nozzle variable spray[J]. Beijing: Journal of Machinery Industry Press, 2013.
[6] Xie F Q, Zhang S F, Gu Z L. Principle and analysis of PSF-50 type feedback flow control nozzle[J]. Journal of Jiangsu Institute of Technology, 1983,4(2): 13-20.
[7] Yang S T, Xie F Q. Study on the step stability of PSF-50 type nozzle[J]. Drainage and Irrigation Machinery, 1984,2(2): 16-18.
[8] Han X Y. Development of a double hit synchronous complete jet nozzle[J]. Water saving irrigation, 1991,2: 38-40.
[9] Han X Y, Sun C H, Shen F X. Double hit synchronous complete jet nozzle [P]. China: 90200784.
[10] Li J S, Kawano H. Sprinkler performances as function of nozzle geometrical parameters[J]. Journal of Irrigation and Drainage Engineering, 1996,122(4): 244-247.
[11] Li J S, Kawano H. Sprinkler performance as affected by nozzle inner contraction angle[J]. Irrigation Science, 1998,18: 63-66.
[12] Tang Y, Yuan S Q, Li H. Automatic test of sprinkler water distribution based on distributed bus[J]. Journal of Agricultural Engineering, 2006,22(4): 112-115.
[13] Ren L Q. Optimization design and analysis of experiments[M]. Beijing: Journal of Higher Education Press,2003.
[14] Robert T P, Victor P. Irrigation process optimization using taguchi orthogonal experiments[J]. Computer and Industrial Engineering, 1998,35(1/2): 209-212.
[15] Lamaddalena N, Fratino U, Daccache A. On-farm sprinkler irrigation performance as affected by the distribution system [J]. Biosystems Engineering, 2007,96(1): 99-109.
[16] Yuan S Q, Li H, Shi W D. Design theory and technology of new spray irrigation equipment[M]. Beijing: China Machine Press,2011.
[17] Li S Y. Theory and design of spray sprinklers[M]. Beijing: Weaponry Industry Press,1995.
[18] Liu J P, Yuan S Q, Li H.Analysis and test of the influence factors on the range and spray uniformity of the complete jet nozzle [J]. Journal of Agricultural Machinery, 2008,39(11): 57-61.
[19] Yuan S Q, Zhu X Y, Li H. The influences of the important structural parameters on the hydraulic performance of the complete jet nozzle[J]. Journal of Agricultural Engineering, 2006,22(10): 113-116.
[20] Sun X. MATLAB 7.0 Basic course[M]. Beijing: Journal of Tsinghua University,2005.
[21] Liu J P. Theory and numerical simulation and experimental study of variable sprinkler[D]. Zhenjiang: Jiangsu University,2008.

[22] Zhu X Y. The theory of complete jet sprinkler and the key technology of accurate sprinkler irrigation[D]. Zhenjiang: Jiangsu University, 2008.

[23] Chen C. Experimental study on stability reliability of complete jet nozzle[D]. Zhenjiang: Jiangsu University, 2007.

[24] Keller J, Bliesner R D. Sprinkle and Trickle Irrigation[M]. New Tork: Van Nostrand Reinhold, 1990.

[25] Liu X L, Niu W Q, Wu P T. Development and hydraulic performance test research of pressure regulator of drip irrigation system[J]. Journal of Agricultural Engineering, 2005, 21 (S): 96 - 99.

[26] Tian J X, Gong S H, Li G Y. Study on the influence of the parameters of the Micro-irrigation pressure regulator on the preset pressure of the exit[J]. Journal of Agricultural Engineering, 2005, 21 (12): 48 - 51.

[27] Tian J X, Gong S H, Li G Y. Economic analysis of the application of pressure regulator in drip irrigation system[J]. Journal of Irrigation and Drainage, 2004, 23(4): 55 - 57.

[28] Khalil M F, Kassab S Z, Elmiligui A A. Applications of drag-reducing polymers in sprinkler irrigation systems: Sprinkler head performance[J]. Journal of Irrigation and Drainage Engineering ASCE, 2002, 128(3): 147 - 152.

[29] Bahceci I, Tari A F, Dinc N. Performance analysis of collective set-move lateral sprinkler irrigation systems used in central anatolia[J]. Turkish Journal of Agriculture and Forestry, 2008, 32: 435 - 449.

[30] Wang G F, Wang W T, Xu F. The optimum calculation of the combination of sprinkler and the combination spacing of the sprinkler system[J]. Journal of Heilongjiang, 2006, 33(2): 36 - 39.

[31] Li X P. Study on the uniformity of water distribution in sprinkler irrigation system[D]. Wuhan: Wuhan University, 2005.

[32] Nelson Corporation. http://www.nelson irrigation.com.

[33] Playán E, Garrido S, Faci J M. Characterizing pivot sprinklers using an experimental irrigation machine[J]. Agricultural Water Management, 2004, 70(3): 177 - 193.

[34] Xu M. Experimental study on low pressure uniform spraying of rotary nozzle [D]. Zhenjiang: Jiangsu University, 2013.

[35] GB/T 19795.1—2005. Agricultural irrigation equipment-first part of rotary nozzle: requirements for structure and operation[S].

[36] GB/T 19795. 2—2005. Agricultural irrigation equipment-second part of rotary nozzle: uniformity of water distribution and test methods[S].

[37] Yang Y C. Design and test of two way step complete jet nozzle[D]. Zhenjiang: Jiangsu University, 2008.

CHAPTER 3

Small-scale Sprinkler Irrigation System

3.1 Introduction

The specific small-scale sprinkler irrigation system is shown in Figure 3-1, the power of the pump is usually around 0.75 kW to 11 kW (15 hp or 15 horsepower)[1]. It is mainly composed of a motor, pump, pipeline, and sprinklers. A direct drive or a belt drive functions as the transmission between the pump and motor[1]. Mobile pipes are also a part of the system's application. The area on a working location in which a fixed point is chosen for spraying can be accomplished by a single sprinkler or multiple sprinklers. The machine will shift to the next location for the irrigation of a larger area when the previous patch of land is watered. It is notable for its small coverage in one irrigation cycle, simple control, and high mobility, and thus it is also referred to as a "small-sized movable sprinkler irrigation system"[2]. Depending on how the system is shifted to the following location, the small-scale sprinkler irrigation system is mainly categorized as a hand-held, hand-lift, trolley-type, and tractor-driven system[3,4]. Concerning different pipe materials, it is divided into systems with an aluminum alloy pipe and those with a plastic-coated hose. The small-scale sprinkler irrigation system is common in China and is also utilized in some developing countries. Figure 3-1 displays the typical configuration of a small-scale sprinkler irrigation system. Figure 3-2 captures jet self-priming irrigation pumps, a more specific mechanic of the overall system. Figure 3-3 and Figure 3-4 focus more closely on jet self-priming irrigation pumps, the first displaying one with a gasoline engine and the latter with a jet self-priming irrigation pump with an electric motor.

Figure 3-1 Small-scale sprinkler irrigation system

Figure 3-2 Jet self-priming irrigation pumps

Figure 3-3 Jet self-priming irrigation pumps with a gasoline engine

Figure 3-4 Jet self-priming pumps with an electric motor

1. Engine

There are three different kinds of engines that can be used in a small-scale sprinkler irrigation system, including the motor, diesel and gasoline.

① Motor engine. A three-phase asynchronous motor with a synchronous speed of 3 000 r/min is the specific type of motor engine that is mainly used. The power equipped in the engine can be 1.1 kW, 1.5 kW, 2.2 kW, 3 kW, 4 kW, 5.5 kW, 7.5 kW or 11 kW. In a portable (or back-pack) sprinkler irrigation system, a single-phase motor is sometimes considered.

② Diesel engine. An air-cooled four-stroke diesel engine with a single cylinder is the most typical diesel engine applied. The power range is generally between 2.2 kW to 11 kW. In the GB/T 25406—2010, the power of a small-scale sprinkler irrigation system is reduced to less than 22 kW[5].

③ Gasoline engine. A small gasoline engine with two strokes at a power range between 0.75 kW to 3 kW and the speed ranges from 5 000 rad/min to 8 000 rad/min is generally used.

2. Pump

A pump lifts water from different natural and man-made water resources (pond, tank or river) to a pressured pipeline to spray it on the land where plants need to be irrigated. By doing this, it ascertains the working pressure and the irrigated area of a system. The technical parameters of a pump include lift head, discharge, power and rotational speed. The common types of pumps include centrifugal pump, self-priming irrigation pump, well pump, and so on.

3. Pipeline

There is a variety of pipelines that can be used for sprinkler irrigation. Light, movable pipelines with a special quick connector attached on each end is a preferable choice, examples being a thinly-walled aluminum alloy pipeline, a galvanized thin-walled steel pipeline, a plastic rigid pipeline, and a flexible plastic-coated pipeline. In small-scale sprinkler irrigation systems, a flexible plastic-coated pipeline is most often utilized. Local conditions such as topography, geology, and climate, as well as the working pressure of a system, combined with the performance and applicable conditions of the said system, are all factors that need to be considered when choosing a type of pipeline.

4. Sprinkler

A sprinkler is also known as a sprayer, which is one of the key components of sprinkler irrigation systems. The structure and hydraulic performance of a sprinkler will largely determine the irrigation quality of a sprinkler irrigation system. With the practice in different areas, the impact sprinkler and the fluidic sprinkler are mainly adopted in small-scale sprinkler irrigation systems. Choosing the correct type of sprinkler for an irrigation system is important in the design process so that it meets the specific necessities that each system requires.

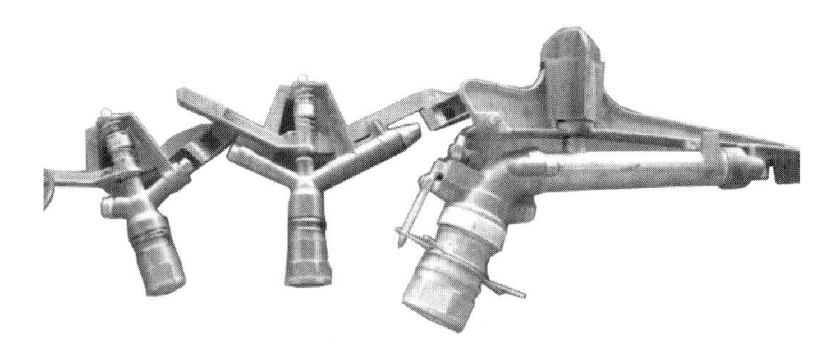

Figure 3-5 PY series impact sprinkler (from left to right: 15PY, 20PY, 30PY.)

3.2 Conventional small-scale sprinkler irrigation system

3.2.1 Small-scale sprinkler irrigation system

Four different types of small-scale sprinkler irrigation systems used depending on the type and quantity of sprinklers equipped: the hand-held gun system (or no spray gun), the single-sprinkler system, the multi-sprinkler system, and the hose-fixed (or semi-fixed) multi-sprinkler system[3,6,7].

① Small-scale sprinkler irrigation system with hand-held guns. The sprayer in this system is hand-held or attached to the end of the pipeline which is directly held when spraying. The atomization index does not need to be high, so the working pressure of the system is low. The system can be applied when the land is dry, and it is easy to move, and thus it is suitable for use in situations that require handling immediately (emergencies) and for drought relief in developing areas. Unfortunately, the sprinkler irrigation uniformity of the system is poor, so it is only practical in more dire situations.

② Small-scale sprinkler irrigation system with single-sprinklers. 40PY or 50PY sprinklers are generally used in the system. This system is unique and notable for its high working pressure, a wide range of spraying, and covering of a large irrigation area by a single sprinkler. However, it has poor irrigation uniformity. The energy consumption in the system is relatively high, but the cost per acre is remarkably low. The coverage of the spraying area for a sprinkler is a fan, so the sprinkler and its stent can easily be moved toward a dry area so that the operation efficiency of the system will be improved. However, the recoil force of the sprinkler when spraying is large and the impact of spraying drops on the land is high, so the soil infiltration capacity has to be considered. The system is suitable for pasture irrigation. This system was popular at the end of the last century, and but its relevance and popularity decrease year by year as technology improves.

③ Small-scale sprinkler irrigation system with multi-sprinklers. Depending on the size of the area and the type of plants, sprinklers with different working pressures and a specific number of sprinklers can be chosen so that it corresponds with the land. Therefore, the parameters of the system such as the flow rate, the pump head, and the power equipped have a wide selection scope, and it has a larger space for the expansion of its use. The irrigation uniformity is higher than those of the two preceding ones, and the energy consumption is reduced.

④ Small-scale sprinkler irrigation system with hose-fixed (or semi-fixed) multi-sprinklers. The pipeline in this system is fixed (or semi-fixed) on the land until an irrigation season is finished and the whole system is packed up and moved to the next area for the irrigation of a different plant or crop. The other parts of the system are very similar to those of the small-scale sprinkling irrigation machine with multi-sprinklers. The operation and management of the small-scale sprinkler irrigation system with hose-fixed/semi-fixed multi-sprinklers are more convenient in comparison to a traditional small-scale sprinkler irrigation system, which is suitable for crops that require seasonally fixed irrigation.

3.2.2 Configuration of the small-scale sprinkler irrigation system

1. Parameters for configuration

The parameters of small-scale sprinkler irrigation systems are shown in Table 3-1[6]. The engine equipped for each system is a diesel engine ranging from 2 to 20 horsepower(1.471 to 14.71 kW) or a motor engine of 1.1 to 18.5 kW. The table covers six types of sprinklers, such as 10PY, 15PY, 20PY, 30PY, 40PY, 50PY and so on[7]. The systems shown are suitable for the irrigation of grain crops and economic crops for a variety of terrain areas.

2. Examples of general configuration of the systems

By using the small-scale sprinkler irrigation systems in Table 3-1, irrigation systems with area of 50～350 acres (3.33～24 hm^2) can be built to fit the scale of farmers' planting land. It has the advantages of convenient operation, low operation cost, and high irrigation uniformity[6,7].

Most of the rural land in China is run by small-scale households and the land is relatively scattered. Therefore, 50 acres (3.33 hm^2) is taken as an example of the configuration in irrigation system. The scheme is presented as follows:

Table 3-1 Configuration parameters of the small-scale sprinkler irrigation systems

System model	Power machine		Pump						Sprinkler					
	Model	Rated power/kW	Pump model	Flow rate/(m^3/h)	Lift/m	Rotation rate/(r/min)	Efficiency/%	Suction/m	Model	Match	Nozzle diameter/mm	Working pressure/MPa	Weted radius/m	Water spray quantity/(m^3/h)
PC20-2.2	R165	2.2	50BP-20	15	20	2 600	58	7~9	10PY	10	5×2	0.15	10	1
PC35-2.9	170F	2.9	50BP-35	15	35	2 600	56	7~9	15PY	7	5×3	0.3	15	2.16
PC35-2.9	170F	2.9	50BP-35	15	35	2 600	56	7~9	20PY	4	6×3.1	0.3	19	2.97
PC35-2.9	170F	2.9	50BP-35	15	35	2 600	56	7~9	30PY	1	12	0.35	27	9.5
PC45-4.4	R175	4.41	50BP-45	20	45	2 600	58	7~9	15PY	9	5×3	0.3	15.5	2.16
PC45-4.4	R175	4.41	50BP-45	20	45	2 600	58	7~9	20PY	6	6×3.1	0.4	19.5	3.4
PC45-4.4	R175	4.41	50BP-45	20	45	2 600	58	7~9	40PY	1	15	0.4	35	17.6
PC55-8.8	S195	8.8	65BP-55	36	55	2 900	64	7~9	20PY	10	6×3.1	0.4	19.5	3.4
PC55-8.8	S195	8.8	65BP-55	36	55	2 900	64	7~9	40PY	2	15	0.35	29.5	14
PC55-8.8	S195	8.8	65BP-55	36	55	2 900	64	7~9	50PY	1	20	0.5	42.3	31.2
PC55-11	S1100	11	80BP-55	50	55	2 900	66	7~9	20PY	15	6×3.1	0.4	19.5	3.4
PC55-13.2	S1110	13.2	CB80-55	50	55	2 900	70	7~9	20PY	18	6×3.1	0.4	19.5	3.4
PC60-18.5	ZS1125	18.5	100BP-60	60	60	2 900	66	7~9	20PY	22	6×3.1	0.4	19.5	3.4

① Scheme 1: the PC30-4.4 hand-held sprinkler irrigation system is selected, shown in Figure 3-6. The basic configuration includes one R165 diesel engine, one 50ZB-30C self-priming pump, one rubber inlet pipe with a diameter of 50 mm and a length of 8 m, and one hand-held frame for the diesel engine.

Option 1 for sprinklers and pipes: one hand-held spray gun, one plastic-coated pipe with a diameter of 50 mm and a length of 20 m;

Option 2 for sprinklers and pipes: seven sets of 15PY impact sprinklers and matching devices (7 × 15PY), one plastic-coated pipe with a diameter of 50 mm and a length of 105 m.

The irrigation area of the system in one specific location is 2.36 acres (0.157 hm^2), the coupling spacing is 15 m×15 m, and the average water application rate is 7.13 mm/h. When the operation time on one location is 2 hours and the irrigation cycle is 5 days, the irrigation area of the system within one irrigation cycle is 47.3 acres (3.153 hm^2).

② Scheme 2: the PC45-4.4 hand-held sprinkler irrigation system is used, shown in Figure 3-7. The basic configuration includes one R175 diesel engine, one 50BP-45 self-priming pump, one rubber inlet pipe with a diameter of 50 mm and a length of 8 m, and one hand-held frame for the diesel engine.

Option 1 for sprinklers and pipes: one spray gun, one plastic-coated pipe with a diameter of 50 mm and a length of 20 m.

Option 2 for sprinklers and pipes: one set of 40PY impact sprinkler and matching devices (1×40PY), one plastic-coated pipe with a diameter of 50 mm and a length of 40 m.

Option 3 for sprinklers and pipes: six sets of 20PY impact sprinkler and matching devices (6×20PY), one plastic-coated pipe with a diameter of 50 mm and a length of 120 m.

The irrigation area of the system of one location is 3.6 acres (0.24 hm^2), the coupling spacing is 20 m×20 m, and the average water application rate is 7.81 mm/h. When the operation time on one location is 2.4 hours and the irrigation cycle is 5 days, the irrigation area of the system within one irrigation cycle is 54 acres (3.6 hm^2).

Figure 3-6 PC30-4.4 hand-held sprinkler irrigation system

Figure 3-7 PC45-4.4 hand-held sprinkler irrigation system

3.3 Movable-fixed double-purpose sprinkler irrigation system

With the development of the rural economy in China, the farmlands inappropriate scale management are gradually increasing. Hence, there is a large demand for the seasonal fixation of an irrigation system. At the same time, some areas where crops, such as corn and wheat, are grown also need fixed irrigation in an irrigation season. In other instances, farmers may hope that the system can be moved conveniently to reduce the investment. Therefore, the movable-fixed double-purpose sprinkler irrigation system needs to be designed, so that the irrigated area of the system can expand and the working mode can be changed or reorganized based on the different necessities of each farmer, and the overall performance of the system can be improved. On the other hand, the installation and moving of a traditional small-scale sprinkler irrigation system are laborious and needs to be addressed.

Through the combination of the working modes of different components in a sprinkler irrigation system, such as movable or fixed, the system can meet the irrigation requirements of different types of farmers and crops. Through the design of the key components which are needed when moving or fixing a system, the adaptability and portability of the system have the potential for improvement.

3.3.1 Movable-fixed double-purpose sprinkler irrigation system

1. Combination mode of the system

A movable-fixed double-purpose sprinkler irrigation system is proposed due to different farmers' needs based on the traditional one-line layout of pipes, as shown in Figure 3-8. According to factors such as topography, water sources, plants, climate conditions, and economic conditions, four specific combination patterns for irrigation systems are constructed for the rural areas in China[6,7].

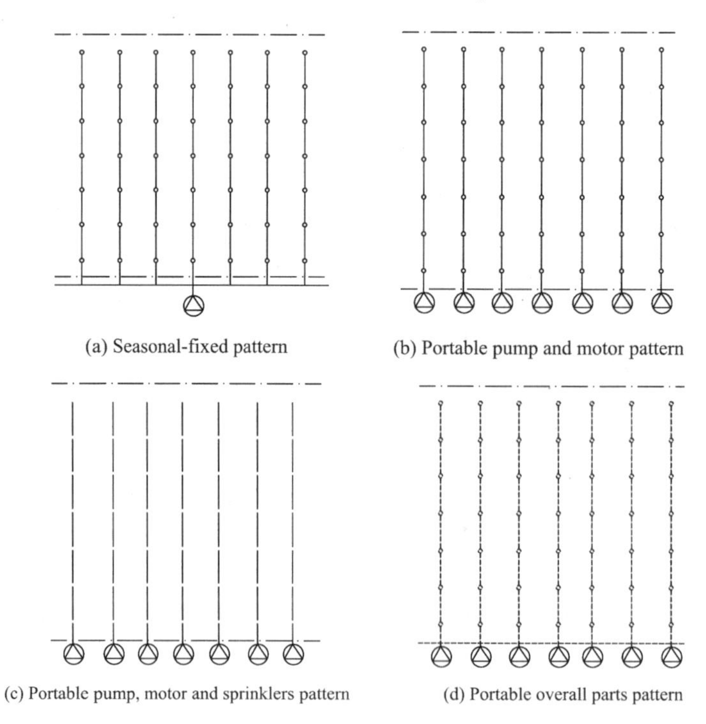

(a) Seasonal-fixed pattern

(b) Portable pump and motor pattern

(c) Portable pump, motor and sprinklers pattern

(d) Portable overall parts pattern

Figure 3-8 Fixed and movable mode of the small-scale sprinkler irrigation system

① Seasonal-fixed pattern. It is suitable for farmers with superior economic conditions, or of the highest quality and a high irrigation frequency, as shown in Figure 3-8a.

② Portable pump and motor pattern. This pattern is suitable for users with most advantageous economic conditions and a relatively high irrigation frequency, as shown in Figure 3-8b. The irrigated areas of the two patterns above are large.

③ Portable pump, motor, and sprinklers pattern. It is suitable for users with decent economic conditions and a relatively low irrigation frequency, as shown in Figure 3-8c. Portable sprinklers would be beneficial to mechanized farming and fixed pipelines would greatly reduce labor intensity in the process of transferring the system.

④ Portable overall parts pattern. It is suitable for users with poor economic conditions and a low irrigation period, as shown in Figure 3-8d.

The combination pattern in the movable-fixed double-purpose sprinkler irrigation system presented is flexible. In the configuration of the system, the existing pump, sprinklers, and plastic-coated pipes are applied so that the versatility of the system is strong and the implementation is convenient. In the field application, the main factors that should be considered in the choice of piping layout of the system including: topographical conditions; water source location; tillage and planting direction; wind direction and wind speed[8].

2. Improvement of key components of the system

Fast-connection pipes were developed as shown in Figure 3-9 to meet the fast-connection necessities of buried submain pipes and movable branch pipes in semi-fixed systems[6]. They can improve the convenience of the connection process and achieve movable-fixed double-purpose sprinkler irrigation.

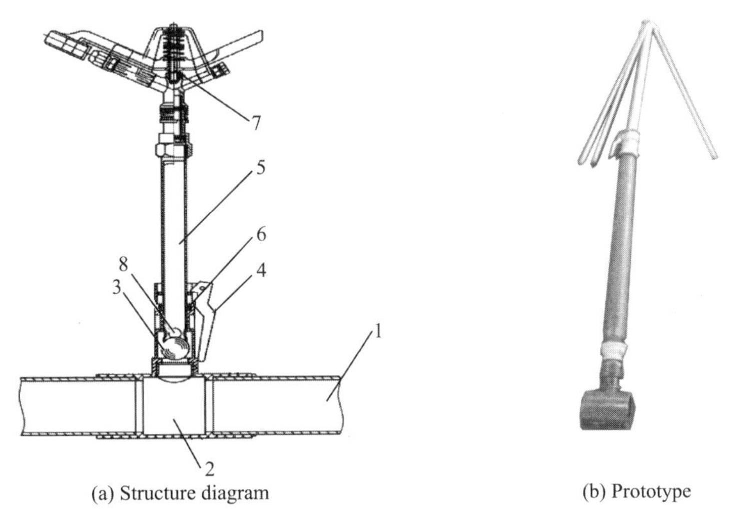

(a) Structure diagram (b) Prototype

1—Submain pipe; 2—Fast three-way joint; 3—Float ball; 4—Handle; 5—Riser; 6—Seal ring; 7—Sprinkler; 8—Inlet of riser

Figure 3-9 Fast-connection pipe

3.3.2 Sprinkler irrigation system with combined double branch pipes

1. System composition

A sprinkler irrigation system with combined double branch pipes is shown in Figure 3-10. It is different and evolved from the traditional one-line layout to a system composed of a submain pipe and many pairs of symmetrical double branch pipes and sprinklers[7]. As shown in Figure 3-11, the system can spray in a rectangular and triangular motion by rotating all the symmetrical branch pipes and sprinklers at a certain angle based on the requirements of the specific irrigation uniformity, shape of the land, direction of the wind and wind speed. In areas with poor economic conditions or when emergency drought relief is needed, the branch pipes can be hand-held directly for irrigation, shown in Figure 3-12.

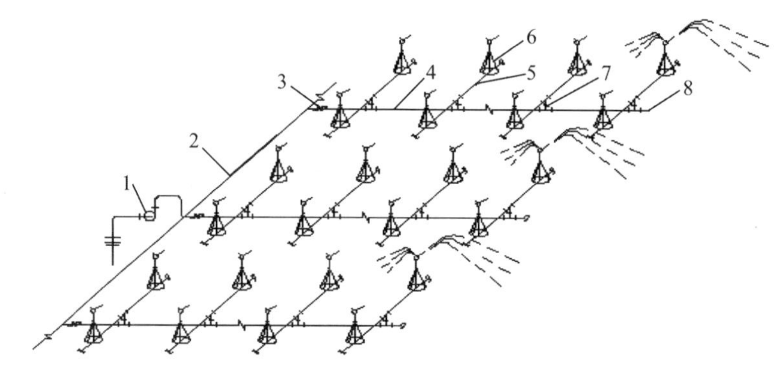

1—Pump system; 2—Water supply pipe; 3—Branch control valve; 4—Branch pipe; 5—Several groups of double-branch pipes; 6—Sprinklers; 7—Combination four-way; 8—Plug

Figure 3-10 A schematic diagram of the distribution structure of a sprinkler irrigation system with combined double branch pipes[7]

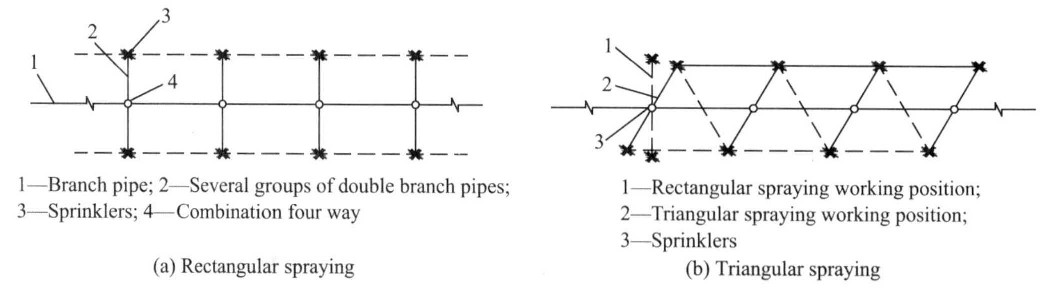

1—Branch pipe; 2—Several groups of double branch pipes; 3—Sprinklers; 4—Combination four way

(a) Rectangular spraying

1—Rectangular spraying working position; 2—Triangular spraying working position; 3—Sprinklers

(b) Triangular spraying

Figure 3-11 Rectangular and triangular spraying

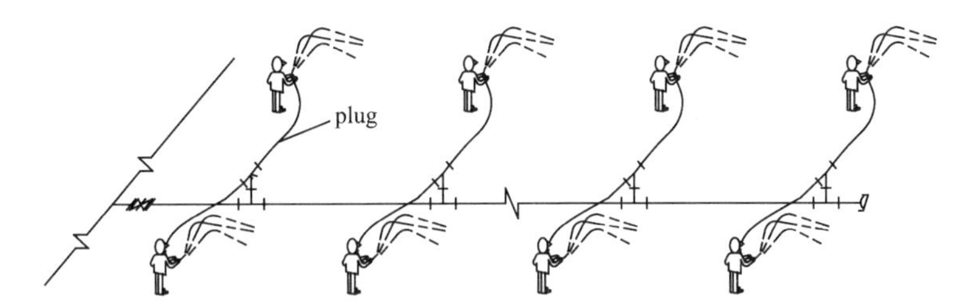

Figure 3-12 The branch pipe holding to water

2. Design of the key components

The key component of a sprinkler irrigation system with combined double branch pipes is the combined cross between the submain pipe and the symmetrical double branch pipes, as shown in Figure 3-13 and Figure 3-14[6]. The upper tee and lower tee junctions make up the combined cross. The upper tee junction is connected to the symmetrical double branch pipes and the lower tee junction to the submain pipe. The sprinkler irrigation system with combined double branch pipes has a smaller pipe diameter, lower cost, and lower energy consumption compared to a traditional small-scale sprinkler irrigation system.

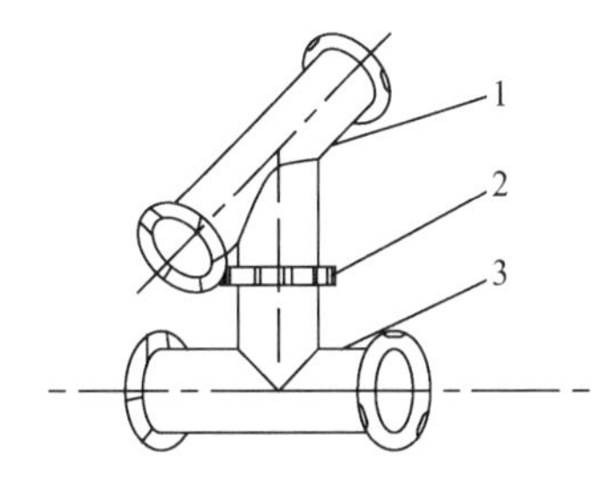

1—Upper tee; 2—Connection; 3—Lower tee

Figure 3-13 Structure diagram of the combined cross

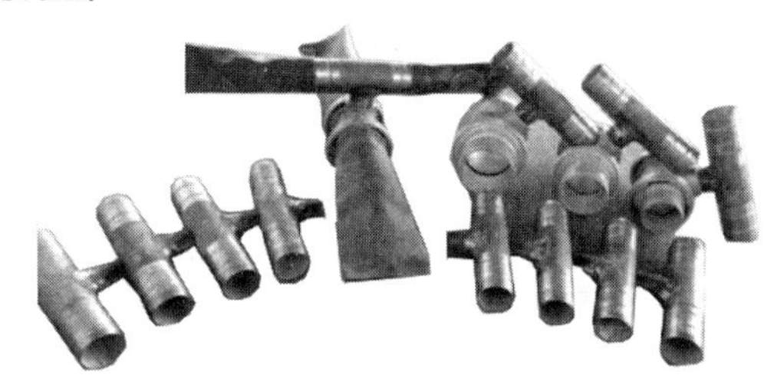

Figure 3-14 The map of the cross

3.4 Optimization design of the small-scale sprinkler irrigation system

Different small-scale sprinkler irrigation systems can be adapted depending on the specificities of any application. Even for the same system, different sprinklers can be configured. Therefore, the comprehensive evaluation of sprinkler irrigation systems is critical to achieving the best results during irrigation.

3.4.1 Evaluation criteria of the small-scale sprinkler irrigation system

1. Irrigation Uniformity

Irrigation uniformity refers to how water is consistently distributed in an irrigated area[6]. It is one of the important indexes of measuring the quality of sprinkler irrigation, which is directly related to crop yield. Irrigation uniformity usually evaluates the coupling uniformity of multiple sprinklers irrigating the same land. Christiansen's uniformity coefficient (C_u) is used to represent the uniformity coefficient, as shown in Eq.(2-11)[8].

2. Specific energy consumption of the system

The total energy consumption of a sprinkler irrigation system operated by T hours is E[9],

$$E=\frac{QHT}{367\eta_b\eta_d}$$
$$T=\frac{m}{1\ 000}\times\frac{S\times10\ 000}{Q\eta_p}=\frac{10mS}{Q\eta_p} \tag{3-1}$$

For a sprinkler irrigation system with N_p locations of spraying, $S=N_pA/2$, $A=2(n-1)ab/10\ 000$, the total energy consumption of a small-scale sprinkler irrigation system can be written as:

$$W_p=\frac{QH}{367\eta_b\eta_d}\times\frac{10mN_pA}{2Q\eta_p}=\frac{N_pmAH}{73.4\eta_b\eta_d\eta_p} \tag{3-2}$$

where E is the energy consumption of the system in kW · h, and T is the running time of the system in h, S is the irrigated area of the system during an irrigation cycle in hm^2, m is the irrigation quota (or the required irrigation depth) in mm, η_b is the pump efficiency, η_d is the transmission efficiency of the motor, η_p is the water efficiency of the application, N_p is the number of working locations of the system, A is the irrigation area of the system at a working location in hm^2, a is the spacing between each sprinkler in m, and b is the spacing between the adjacent working locations in m.

Eq.(3-2) shows the direct correlation between the total energy consumption of the sprinkler irrigation system and the irrigation area S and quota m. Therefore, the energy consumption by the sprinkler irrigation system can be defined as the energy consumed per depth of water sprayed over a hectare, which can be obtained by Eq.(3-2)[9],

$$E_p=\frac{QH}{367\eta_b\eta_d}\times\frac{10\times1\times1}{Q\eta_p}=\frac{H}{36.7\eta_b\eta_d\eta_p} \tag{3-3}$$

where E_p is the specific energy consumption of the system in kW · h/(mm · hm^2).

With Eq.(3-3), the efficiency and the pump head are the only givens needed to evaluate and analyze the energy consumption of a sprinkler irrigation system. The formula is concise and convenient in the calculation, and it satisfies the requirements of evaluation criteria. Moreover, the specific energy consumption of the system as an evaluation criterion may reflect the running characteristics of the sprinkler irrigation system.

3. Annual cost of the sprinkler irrigation system

Different calculation methods depend on the scale and characteristics of a specific sprinkler irrigation project. The annual cost of a small-scale sprinkler irrigation system is mainly comprised of the specific depreciation expense and specific energy consumption fee. This assumes the cost without the water fee and the application of the static depreciation method. The annual cost of the system can be derived as follows[10]:

$$C_A = \frac{r(C_b + naC_g + nC_s)}{MA} + \frac{ETQH}{367.2\eta_b \eta_d \eta_p MA} \tag{3-4}$$

where C_A is the annual cost per hectare of land irrigated using the system in RMB/(a • hm^2), C_b is cost of power machine, water pump and water inlet pipe, in RMB/set, C_g is unit price of the pipe in RMB/m, C_s is unit price of the sprinkler, upright tube and joint in RMB/set, E is the fuel price, converted from the corresponding average fuel price on the market, fuel density, and calorific value in RMB/(kW • h), T is the annual running time of the system in h. In Eq.(3-4), the first value is the specific depreciation expense in RMB/(a • hm^2), and the second value is the specific energy consumption fee in RMB/(a • hm^2)[6, 10].

3.4.2 Hydraulic calculation method for the design of pipelines in a small-scale sprinkler irrigation system

The hydraulic calculation methods for the irrigation pipelines mainly utilize the forward method and the backward method. The forward method is used for the preliminary design of the system when the performance parameters of the pump are known and the working pressure of sprinkler at the end of the pipe has been checked. Conversely, the backward method is used when the operating condition of the system is unknown and the working pressure of the sprinkler at the end of the pipe needs to be confirmed.

1. Stepwise hydraulic calculation with the forward method[11]

Figure 3-15 is the layout of a small-scale sprinkler irrigation system. The main components include the pump system, the pipeline, and the sprinklers. As shown in the figure, the diameter of the pipe is d, and n represents the number of sprinklers on the pipeline. The distance between the first sprinkler and the pump is represented as l_0, and the distance between the two adjacent sprinklers l. For an entire riser to be simplified into one sprinkler, it is assumed that all risers and sprinklers are the same model and size.

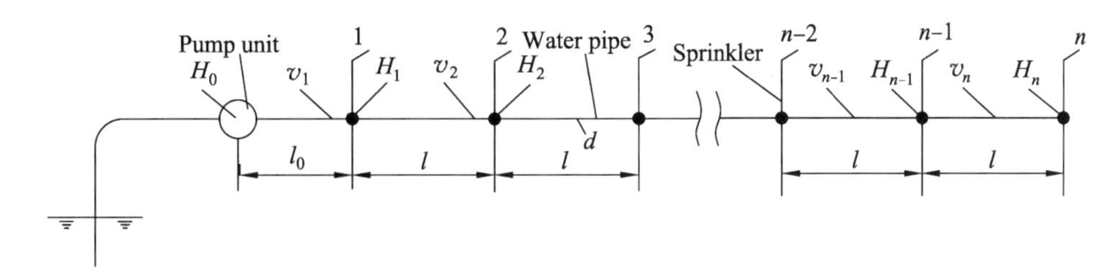

Figure 3-15 The layout of as mall-sized light sprinkler irrigation system

Since every sprinkler represents its corresponding riser, the calculation of the head loss contains the local head loss at the tee and the frictional head loss on the riser. Therefore, the total loss between any two sprinklers can be calculated by the frictional head loss along the pipeline connecting two sprinklers multiplied by an experience coefficient. The Darcy Weiss Bach formula is used to analyze the water delivery pipeline of the sprinkler irrigation system step by step. Suppose the working pressure water column of the sprinkler 1, 2,…, n are H_1, H_2,…, H_n.

If a sprinkler irrigation system utilizes hydraulically smooth pipelines, the calculation of λ can be obtained by the Blasius Formula.

$$\lambda=0.3164\left(\frac{dv}{\upsilon}\right)^{-0.25} \tag{3-5}$$

where υ is the fluid kinematic viscosity in m^2/s.

For the case of a circular tube, $d=4R$. The working pressure head of the first sprinkler can be determined by

$$H_1=H_0-k_0\lambda_0\frac{l_0}{d}\frac{v_1^2}{2g} \tag{3-6}$$

where H_0 is pressure head of the pump outlet in m, k_0 is the experience coefficient, λ_0 is the drag coefficient of the first section of the pipeline, d is the internal diameter of the pipe in m, and v_1 is the flow rate in the first section of the pipe in m/s.

The working pressure head of the second sprinkler can be determined as follows:

$$H_2=H_1-k_1\lambda_1\frac{l}{d}\frac{v_2^2}{2g} \tag{3-7}$$

In the same way, the pressure head of the third sprinklers can be obtained:

$$H_3=H_2-k_2\lambda_2\frac{l}{d}\frac{v_3^2}{2g}=H_1-\frac{k_1\lambda_1 lv_2^2}{2gd}-\frac{k_2\lambda_2 lv_3^2}{2gd} \tag{3-8}$$

The working pressure of the n-th sprinkler can be obtained:

$$H_n=H_1-\frac{k_1\lambda_1 lv_2^2+k_2\lambda_2 lv_3^2+\cdots+k_{n-1}\lambda_{n-1}lv_n^2}{2gd} \tag{3-9}$$

In summary,

$$H_n=H_1-\sum_{i=1}^{n-1}\frac{k_i\lambda_i lv_{i+1}^2}{2gd} \tag{3-10}$$

Substituting Eq.(3-6) in Eq.(3-11),

$$H_n=H_1-\sum_{i=1}^{n-1}\frac{0.316\,4lk_iv_{i+1}^2}{2gd}\times\left(\frac{dv_{i+1}}{\upsilon}\right)^{-0.25} \tag{3-11}$$

Thus,

$$H_n=H_1-\frac{0.316\,4l}{2gd}\times\left(\frac{\upsilon}{d}\right)^{0.25}\sum_{i=1}^{n-1}k_iv_{i+1}^{1.75} \tag{3-12}$$

2. Backward hydraulic calculation iteration method

To calculate the energy consumption of a sprinkler irrigation system, the operating conditions of the system must first be determined, which is the intersection point of the characteristic curves of a pipeline and pump. The characteristic curve of the pump is usually issued by the manufacturer, but in any case, the pipeline characteristic curve can be obtained through a hydraulic calculation[12, 13].

The pipeline is numbered in the form of Figure 3-16, and the hydraulic calculation is implemented according to the following steps and formulas.

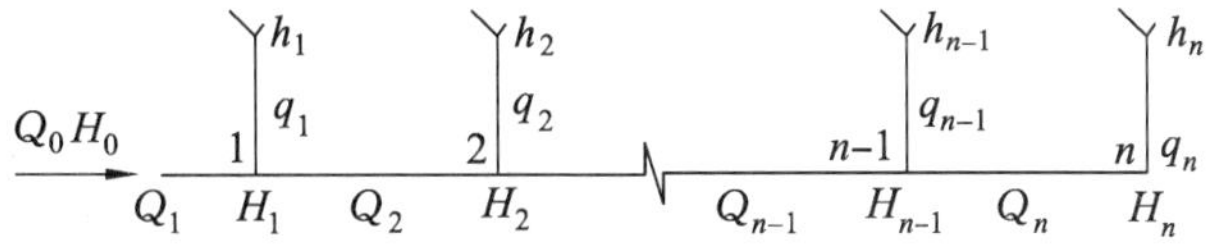

Figure 3-16 Layout of the pipeline in a small-scale sprinkler irrigation system

① Assuming a series of values for the pressure head of the end sprinkler h_n, calculate the flow rate of the riser and the pressure head and flow rate at the end of the delivery pipeline:

$$q_n = \mu \frac{\pi d_p^2}{4} \sqrt{2gh_n} = 0.012\ 52 \mu d_p^2 h_n^{0.5} \tag{3-13}$$

$$H_n = h_n + f \frac{q_n^m}{d^b}(l + le_n) + l \tag{3-14}$$

$$Q_n = q_n \tag{3-15}$$

② Calculate the pressure head and discharge of the sprinklers and pipeline from Section $n-1$ to Section 1:

$$H_i = H_{i+1} + f \frac{Q_{i+1}^m}{D_{i+1}^b}(a + Le_{i+1}) + aI_i \tag{3-16}$$

$$h_i = H_i - f \frac{q_i^m}{d^b}(l + le_i) - l \tag{3-17}$$

$$q_i = \mu \frac{\pi d_p^2}{4} \sqrt{2gh_i} = 0.012\ 52 \mu d_p^2 h_i^{0.5} \tag{3-18}$$

Eq.(3-19) and Eq.(3-22) will be solved using the iterative method:

$$Q_i = Q_{i+1} + q_i \tag{3-19}$$

③ Calculate the pressure head and flow rate at the inlet of the pipeline:

$$H_0 = H_1 + f \frac{Q_1^m}{D_1^b}(a + Le_1) + aI_1 \tag{3-20}$$

$$Q_0 = Q_1 \tag{3-21}$$

where n is the number of sprinklers, H_i is the pressure head of the pipe at Node i along the delivery pipeline in m(H_2O), Q_i is the flow rate at Pipe Section i in m^3/h, H_1 is the inlet pressure head of the pipe in m(H_2O), Q_0 is the inlet flow rate of the pipe in m^3/h, h_i is the sprinkler pressure head at Node i in m(H_2O), q_i is the sprinkler flow rate at Node i in m^3/h, a is the spacing of sprinklers in m, l is the lengths of the riser in m, D_i and d are the internal diameters of the pipeline and the riser in mm, Le_i and le_i are the equivalent lengths of the local head loss in the main pipe and the riser in m(H_2O), I_i is the terrain slope, μ is the flow rate coefficient of the sprinkler, $\mu =$

$79.87 q_p d_p^{-2} h_p^{-0.5}$, h_p is the rated working pressure head of the sprinkler in m, q_p is the flow rate of the sprinkler in m^3/h, d_p is the nozzle diameter in mm, and f, m, and b are the calculation coefficients of the head loss correlated with the pipe material.

④ After the hydraulic calculation, supposing $Q = Q_0$, the pump working condition can be determined by the head-flow rate curve through the cubic polynomial fitting calculation, as follows:

$$H = c_0 + c_1 Q + c_2 Q^2 + c_3 Q^3 \tag{3-22}$$

where c_0, c_1, c_2, c_3 are regression coefficients of the polynomial fitting equation for the relationship between the head and flow rate, Q is the flow rate of the pump in m^3/h.

In the same way, the pump efficiency is derived by

$$\eta_b = b_1 Q + b_2 Q^2 + b_3 Q^3 \tag{3-23}$$

where b_1, b_2, and b_3 are regression coefficients of the polynomial fitting equation for the relationship between efficiency and discharge.

⑤ Draw the characteristic curve of the pipeline after the relationship between the series of inlet pressure heads and discharges of the pipeline has been obtained.

3.4.3 Optimization model of a small-scale sprinkler irrigation system

In the design of a sprinkler irrigation system, the sprinklers, pipeline, and pipe fittings are equipped according to the capacity of the pump whereas the size of sprinklers is dependent on the soil, plant conditions, and operative environment [14,15]. In this case, the optimization of the system is usually configured based on the lowest possible initial investment, energy consumption, or annual cost. Once this is determined, the optimal number of sprinklers, pipeline diameters, and other working parameters are implemented. The optimization mathematical model of the irrigation system is established based on the layout of the pipeline shown in Figure 3-15.

1. Optimization objective

When the specific pumps and sprinklers are selected, the differences in cost for different configurations of each type of system are mainly determined by different diameters of every pipeline. The pipes for most small-scale sprinkler systems are plastic-coated and inexpensive, and thus various pipe diameters only have a small impact on the entire cost of each system. Therefore, the specific energy consumption is considered and assumed the most influential establishment of the optimization model, as $\min E_p$.

2. Constrained conditions

(1) Minimal working pressure of the sprinkler

According to GB 50085—2007 in China, it is recommended that any sprinkler should work no lower than 90% of the designated pressure so that the wetted radius and flow rate of the sprinkler will not have high variance in the operation.

$$h_{\min} \geqslant 0.9 h_p \tag{3-24}$$

where h_{min} is the minimal working pressure head of a sprinkler in m(H_2O) and h_p is the designed working pressure head of a sprinkler in m(H_2O).

(2) Working pressure deviation of the sprinkler

The current GB 50085—2007 specifies that the pressure difference between any two sprinklers along the same branch should be lower than 20% of the originally designated pressure of the sprinkler to ensure a generally equal application of irrigation uniformity, so

$$h_v = \frac{h_{max} - h_{min}}{h_p} < 20\% \tag{3-25}$$

where h_v is the working pressure deviation rate of a sprinkler, and h_{max} and h_{min}, are the maximum and minimum working pressure of a sprinkler in m.

(3) Pump and pipeline working conditions

The intersection point between the pipeline characteristic curve and the pump flow rate-head curve is the actual operating point of the system. In the optimization process, consistency between the pump working condition and the pipeline working condition must be ensured. Pipeline characteristics are determined by the hydraulic calculation of the pipeline, considering the calculation accuracy and simplifying the calculation requirements. The characteristic curve of the pipeline is gathered through the hydraulic calculation, while the working condition of the pump can be found when studying the flow rate-head curve through the cubic polynomial fitting method. This method is utilized for calculation accuracy and simplicity, as shown by Eq.(3-22).

In the optimizing process, the flow rate of the pump is equal to the pipe inlet flow rate Q_0 through the hydraulic calculation of the pipeline and the pump head shown in Eq.(3-22). The pump-pipeline condition can only be satisfied when the pump head is following the inlet pressure of the pipeline.

$$H_0 = H \tag{3-26}$$

where H_0 is the inlet pressure of the pipeline obtained by hydraulic calculation in m.

(4) The number of sprinklers

The following equation determines the maximum amount of sprinklers on the main pipe that have consistent diameters, outlet spacing, and flow rate.

$$N_m = \mathrm{INT}\left\{\left[\frac{(m+1)[\Delta h]D^b}{kfaq_p^m}\right]^{\frac{1}{m+1}} + 0.52\right\} \tag{3-27}$$

Then the range of the number of sprinklers on the level pipeline as follows:

$$N_m(D_{min}) \leqslant n \leqslant N_m(D_{max}) \tag{3-28}$$

where N_m is the maximum amount of sprinklers, n is the actual number of sprinklers, $[\Delta h]$ is the minimum pressure of sprinklers, $[\Delta h] = 0.2h_p$, D is the pipe diameter in mm, a is the sprinkler spacing in m, k is the coefficient of local water head loss, $k = 1.1 \sim 1.15$, f, m, and b are calculation coefficients of head loss in relation to pipe material, q_p is the designed flow of sprinkler in m^3/h, and D_{max} and D_{min} are the maximum and minimum of the internal diameters of the alternative pipeline in mm.

References

[1] Pan Z Y, Liu J R, Shi W D. Status of light weight and small size movable sprinkler irrigation set and its gap with advanced countries[J]. Drainage and Irrigation Machinery, 2003,21 (1): 25 - 28.

[2] GB/T 25407—2010. Series small-sized light movable irrigation machines[S].

[3] Yuan S Q, Li H, Wang X K. Status, problems, trends and suggestions for water-saving irrigation equipment in China [J]. Journal of Drainage and Irrigation Machinery Engineering, 2015,33(1): 78 - 92.

[4] Tu Q. Study on evaluation system of the energy consumption in the small-scale sprinkler irrigation machine[D]. Zhenjiang: Jiangsu University,2011.

[5] GB/T 25406 - 2010. Series small-sized light sprinkler irrigation systems[S].

[6] Tu Q. Optimization design and experimental study of small-scale sprinkler irrigation systems with low energy consumption and multi purposes[M]. Zhenjiang: Jiangsu University,2014.

[7] Tu Q, Li H, Wang X K. Optimization design and experimental study of small-scale sprinkler irrigation systems with low energy consumption and multi purposes[M]. Beijing: Science Press,2017.

[8] GB 50085—2007. Technical code for sprinkler engineering[S].

[9] Wang X K, Yuan S Q, Liu J R. Energy consumption calculation and evaluation methods for light and small unit sprinkler system[J]. Journal of Drainage and Irrigation Machinery Engineer,2010,28(3): 247 - 250.

[10] Tu Q, Li H, Wang X K. Configuration and design optimization of small-scale sprinkler irrigation systems based on different indicators [J]. Transactions of the Chinese Society of Agricultural Engineering, 2013, 29 (22): 83 - 89.

[11] Zhu X Y, Cai B, Tu Q. Head loss hydraulic calculation step by step for light and small sprinkler irrigation system[J]. Journal of Drainage and Irrigation Machinery Engineering, 2011,29(2): 180 - 184.

[12] Wang X K, Yuan S Q, Zhu X Y. Optimization of light-small movable unit sprinkler system using genetic algorithms based on energy consumption indicators [J]. Transactions of the Chinese Society for Agricultural Machinery, 2010,40(10):58 - 62.

[13] Tu Q, Wang X K, Li H. Optimization of sprinkler irrigation machine based on genetic algorithms[C]. ASABE Annual International Meeting, Dallas, Texas, USA, 2012.

[14] Tu Q, Li H, Wang X K. Ant Colony Optimization for the design of small-scale irrigation systems[J]. Water Resources Management, 2015, 29 (7): 2323 - 2339.

[15] Tu Q, Li H, Wang X K. Multi-criteria evaluation of small-scale sprinkler irrigation systems using grey relational analysis [J]. Water Resources Management, 2014,28(13): 4665 - 4684.

CHAPTER 4

Large and Medium-sized Sprinkler Irrigation System

There is no international standard for a large or medium-sized sprinkler irrigation system, but this label usually means the reel type sprinkler irrigation system, the side-roll wheel sprinkler irrigation system, the center pivot sprinkler irrigation system, and the lateral move sprinkler irrigation system.

4.1 Introduction

In the 1920s, in order to reduce labor intensity, prevent laborers from having to maneuver heavy mobile pipeline irrigation systems, and better understand the mechanization of farmland irrigation, side-roll wheel sprinkler irrigation systems were successively developed in the former Soviet Union and the United States[1]. It was firstly promoted in the United States in the year 1950. By 1982, the total U.S. sprinkler irrigation area using side-roll wheel sprinkler irrigation systems had reached 220 million hectares nationwide, accounting for 16.9% of all sprinkler irrigation areas in the country. From 1960 to 1985, side-roll wheel sprinkler irrigation systems had also been widely used in the former Soviet Union and Eastern European countries[2]. Side-roll wheel sprinkler irrigation systems were developed based on previous sprinkler irrigation systems involving artificially transportable pipelines. The early side-roll wheel sprinkler irrigation system was moved by an artificial pushing roller, which reduced labor intensity to a large extent. The transfer method (manual labor) was gradually replaced by power machines, further reducing labor intensity. Side-roll wheel sprinkler irrigation systems are suitable for irrigating almost all crops, including mollugo, Chinese herbal medicine, cotton, vegetables, wheat, cereal, soybeans, sugar beets, potatoes, and so on. However, the side-roll wheel sprinkler irrigation system is not suitable for high stalk crops such as corn or sugarcane, due to the height limitation of the pipeline (axle); when the crop grows past a certain height, the axle will not be able to reach the crop.

By the early 1950s, the United States had also developed a center pivot sprinkler irrigation system. The center pivot sprinkler irrigation system was firstly established in the United States in 1952. In the early days, the hydraulic drive was the main feature of the center pivot sprinkler irrigation system, but then an electric drive center pivot sprinkler irrigation system was developed in 1965. Thereafter, the center pivot sprinkler irrigation system was widely used internationally. In the United States, the sprinkler irrigation area of the center pivot sprinkler irrigation system accounted for 40% of the total irrigated area in the nation by the 1970s. Once the center pivot sprinkler irrigation system was praised by the famous scientific journal "Science America" as "the most significant agricultural machinery invention since the tractor replaced the livestock", igniting its wide usage. There are many overseas manufacturers of center pivot sprinkler irrigation systems, such as Lindsay, Valmont, Reinke, TL, and Lockwood in the United States, Bauer in Austria, Irrifrance in France, RKD in Spain, Ireland in Italy, and AJkhoryaef in Saudi Arabia. The center pivot sprinkler irrigation system is suitable for almost all kinds of soil texture and field crops, cash crops, vegetables, and herbage. It can also be used in contiguous cropland with no resistance from the ground.

However, unlike the development of center pivot sprinkler irrigation system in the United States, a reel type sprinkler irrigation system was designed and developed in Europe in the 1960s, By the late 1970s, European countries such as France, Germany, Italy, Austria, Sweden, Denmark, Spain, and the United Kingdom had finalized and mass-produced their products, which appeared many manufacturing enterprises such as IRRIFRANCE in France and BAUER in Austria. Italy specifically became the gathering place of global reel-type sprinkler irrigation system manufacturers, as there were more than a dozen irrigation manufacturers there. At present in Italy, after more than 40 years of development, several world-renowned enterprises, with successful brands and technologies have formed, such as OCMIS, MARANI, RM, CASELLA, IRRIMEC, and IRRILAND. The total reel type sprinkler irrigation systems in Italy reach an annual output ranging from 6 000 to 7 000 units and hold a total capacity of 150 000 units, covering dozens of series reel type sprinkler irrigation systems. Today, China has also become a largely successful manufacturing country for reel type sprinkler irrigation systems, and its output is predicted soon to be the largest in the world. Well-known enterprises include Jiangsu Huayuan, Jiangsu Xinge, and Hebei Nonghaha. In comparison to foreign countries, research in China is more open and beneficial. In particular, the solar-powered reel type sprinkler irrigation systems have the advantages of low energy consumption and highly controlled precision of moving speed compared to overseas hydraulically driven reel type sprinkler irrigation systems[3-9]. The reel type sprinkler irrigation system can adapt to all kinds of lands of different sizes, shapes and terrain gradients. It can also irrigate a diverse spectrum of crops, such as corn, soybeans, potatoes, herbage, fruits, and cash crops (such as sugarcane, tea leaves, bananas, etc.).

The lateral move sprinkler irrigation system was developed in the 1970s. It was designed to overcome the lack of effective irrigation by the center pivot sprinkler irrigation system, as nearly 22% of the square area could not be covered. In 1971, Wade Rain Company in the United States developed the cable-drive, hydrodynamic lateral move sprinkler irrigation system based on the working principle of the cable-driven reel type sprinkler irrigation system, which drags the hose as it moves forward. In 1977, the Vamoni Company in the United States developed an electric lateral move sprinkler irrigation system based on the center pivot sprinkler irrigation system, taking advantage of the buried guide tracking feature. Around the same time, the Lindsay Company in the United States developed an electric lateral move sprinkler irrigation system with fixed cable guidance along canal floors[10].

At present, various types of sprinkler irrigation machines have irrigated tens of millions of hectares of arable land, dunes, and grasslands around the world. This is a revolutionary achievement for the international history of agricultural irrigation.

4.2 Reel type sprinkler irrigation system

4.2.1 Structure and working principle

A reel type sprinkler irrigation system is mainly composed of a sprinkler cart, water supply hose (PE hose), reel, reel driver (water turbine), and gearbox, as shown in Figure 4-1.

Figure 4-1 The reel type sprinkler irrigation system

The reel type sprinkler irrigation system is fixed through screwing its ground anchor into the soil after it is driven by a power machine to a field. The rotation of the water turbine is driven by pressured water diversion from the pump station flowing into the water-turbine case by the water inlet. Then, the pressured water flows into the hollow shaft of the reel center and passes through the PE reel pipe coiled around the reel frame along the radial direction. Finally, the water is sprayed from the rotating sprinkler installed in the sprinkler cart at the bottom of the PE pipe. The PE pipe is stretched to the most distant point of the fully deployed pipe away from the reel frame with the sprinkler cart running by the dragger during its initial state. When the sprinkler irrigation machine is stationary, the reel is rotated by a speed-changing mechanism, such as a gear reducer, under the drive of a water turbine (or other power sources). The power source constantly pulls the PE pipe by cycling and winding the reel frame. During the recovery process, the sprinkler continuously sprays water in the field. When the PE pipe is wound to the head, the sprinkler cart is automatically lifted to a transporting position by the automatic lifting frame and collides with the closing rod to decelerate the gearbox. Eventually, the reel stand rotation stops automatically, ending the operation[11].

4.2.2 Performances and characteristics

The critical factor influencing the effectiveness of the sprayed water and the mechanical performance of a reel type sprinkler irrigation system is the traveling speed of the sprinkler cart. Many factors can cause the change of the traveling speed of a sprinkler cart, the most important of which are the following two aspects[12]:

(1) Changes of the hose diameter on a reel

Changes and adjustments for spray uniformity are shown in Figure 4-2. At the start of irrigation on each hose strip, the sprinkler cart lies at the end of the hose strip, and the hose is almost fully spread across the ground. The nearer the sprinkler cart to the reel, the shorter the length of the drawn PE hose; the more the PE hose is coiled on the reel, the larger the coiling diameter of PE hose. If the rotational angular velocity of reel remains constant, the traveling speed of the sprinkler cart v will gradually speed up and equal to v_1, resulting in a gradually decreasing water supply amount h on the ground, equalling to h_1.

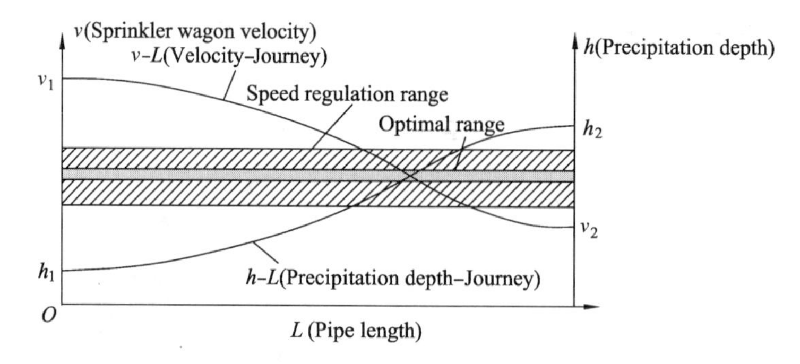

Figure 4-2 Variation and adjustment of the speed and water supply

(2) Changes of the hose (including internal water) quality on a reel

The quality of the reel will improve as the number of layers coiling the reel increases, and the friction between hose and ground will gradually decrease as the hose becomes more fully coiled on the reel from previously being spread across the ground at the beginning of the process. The traveling speed v of the sprinkler cart with a single nozzle will become more rapid and gradually equals to v_1 if the drive power remains constant, so the amount of water supply h on the ground decreases and gradually equals to h_1.

4.2.3 Field design

1. Layout

When using a reel type sprinkler irrigation system for sprinkler irrigation, the irrigation area should be laid out in long strip plots. The plot size should be calculated based on the length of the hose and the wetted radius of the sprinkler (considering the overlaps on both sides). A pathway in the irrigation area should be created, so the sprinkler irrigation machine can be transported; the path should be perpendicular to the crop planting line, as shown in Figure 4-3.

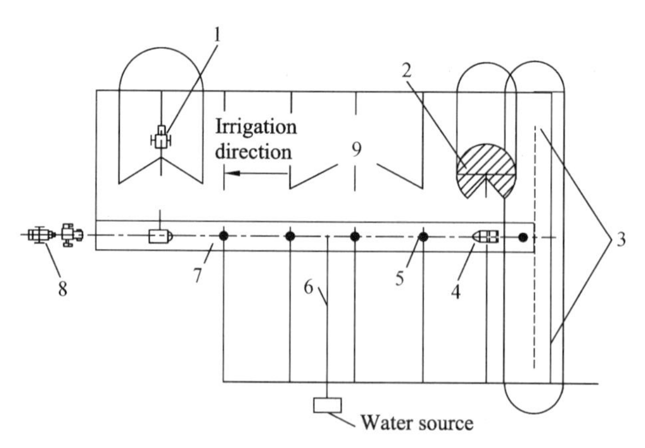

1—Sprinkler trolley pulled by tractor to the strip; 2—Sprinkler trolley; 3—Irrigated strip; 4—Reel truck; 5—Hydrant; 6—Water pipe; 7—Passage for reel truck; 8—Irrigator pulled by tractor to the plot

Figure 4-3 Field layout for the reel type sprinkler irrigation system

When using pipelines for water supply, it is the most effective to arrange the water supply pipeline along the direction of the roadway, to set the hydrant at the water intake area of the sprinkler irrigation machine. This makes the connection between the sprinkler irrigation machine and hydrant more reliable.

When using a supply from a channel of water, it is advisable to arrange them parallel to the roadway to prevent seepage. When a channel's depth cannot meet the depth of a water pump for the sprinkler irrigation machine, it is recommended to set a working pool at the water intake area of the sprinkler irrigation machine. The size of the work tank should meet the normal water pump and dredging requirements.

2. Design[13]

(1) The maximum irrigation quota can be determined as:

$$m_0 = 0.1\gamma h(\beta_1 - \beta_2) \tag{4-1}$$

where m_0 is the maximum irrigation quota in mm, γ is the soil bulk density in g/cm^3, h is the planned depth of wetting layer in cm, β_1 is the maximum quota of soil moisture content (weight percent), and β_2 is suitable for soil with lower water content (weight percentage).

(2) Design irrigation quota can be determined as:

$$m \leqslant m_0 \tag{4-2}$$

where m is the irrigation quota measured in mm.

(3) Design irrigation cycle can be determined as:

$$T = m/W \tag{4-3}$$

where T is the irrigation cycle in d, and W is the water consumption over a period of a single day in mm/d (it can also be interpreted as the average daily water requirement of the most critical period of crop water usage).

(4) Water flow rate of a controlled irrigation area for a single sprinkler irrigation machine can be calculated as:

$$Q = 10WA/(t\eta_p) \tag{4-4}$$

where Q is water flow rate of a controlled irrigation area by a single sprinkler irrigation machine in m^3/h, A is controlled irrigation area for a single sprinkler irrigation machine in hm^2, t is the designated time of day for irrigation in h, which is suitable for a period ranging from 12 to 20 h, and η_p is the effective coefficient of spray water. When the wind speed is less than 3.4 m/s, η_p is taken as 0.8 to 0.9, and when the wind speed ranges from 3.4 to 5.4 m/s, η_p is taken as 0.7 to 0.8.

(5) The demand flow rate of the sprinkler irrigation machine can be determined as:

$$Q_0 \geqslant Q \tag{4-5}$$

where Q_0 is the demand flow rate of the sprinkler irrigation machine in m^3/h.

(6) The width of irrigation strips can be determined as follows:

① The irrigation strip width for a single sprinkler cart can be calculated as:

$$b = 2kR_1 \tag{4-6}$$

where b is the width of the irrigation strip in m, k is the reduction factor of the wetted radius (as shown in Table 4-1), and R_1 is the wetted radius of the sprinkler in m.

Table 4-1 The range reduction factor

Wind speed/(m/s)	k
0.3～1.6	0.8～0.7
1.6～3.4	0.7～0.6
3.4～5.4	0.6～0.5

② The irrigation strip width for a trussed sprinkler cart can be calculated as:

$$b = B + 2kR_2 \tag{4-7}$$

where B is the truss length in m, and R_2 is the wetted radius of the sprinkler at both ends of the truss in m.

(7) A continuous effective spray length can be calculated as:

$$S = L + 0.5R \tag{4-8}$$

where S is the continuous effective spray length in m, L is the length of the hose spread upon the ground in m, 0.5 is the reduction factor, and R is the wetted radius of the sprinkler in m, it equals to R_1 when sprayed using a single sprinkler cart and equals to R_3 when sprayed using a trussed sprinkler cart; R_3 is the wetted radius of other sprinklers except for the sprinkler at both ends of the truss in m.

(8) Water application rate can be determined by the following methods:

① Experimentation (trial and error);

② Calculating the equation of the water application rate of a single sprinkler cart with no wind condition, represented by Eq.(4-9):

$$\rho = (1\,000 \times 360 \times q)/[\pi \times (0.9 \times R_1)^2 \times \alpha] \tag{4-9}$$

where ρ is the water application rate in mm/h, q is the flow rate of a single sprinkler in m^3/h, α is the spray sector-angle in degrees, which is usually taken as 200° to 300°, and 0.9 is the reduction factor.

(9) The traveling speed of a sprinkler cart at work can be calculated as:

$$v = (1\,000 \times Q_0 \times \eta_p)/(m \times b) \tag{4-10}$$

where v is the traveling speed of a sprinkler cart at work in m/h, and Q_0 is the demand flow rate of a sprinkler irrigation machine in m^3/h.

4.3 Center pivot sprinkler irrigation system

4.3.1 Structure and working principle

A center pivot sprinkler irrigation system connects the end of a long, thinly walled metal pipe with many sprinklers functioning with a rotatable elbow to a fixed standpipe at the center of a plot, with the standpipe acting as a fulcrum shaft and supplying pressurized water. Thinly walled metal pipes lay overhead on several equally spaced tower wagons. The central axis and each tower wagon have a system to ensure the reliability of the pipe's moving so that it will consistently remain in a straight line and conduct sprinkler irrigation at a regulated speed by rotating the center pivot continuously.

At present, center pivot sprinkler irrigation systems function in an electrically driven mode, and a hydraulically driven mode; the most common mode is electrically driven. Electric center pivot sprinkler irrigation systems are composed of center support, tower wagon, end cantilever, and electric control synchronization system. The truss equipped on sprinklers is supported by several tower wagons. The produced center pivot sprinkler irrigation system in domestic and foreign companies has been standardized and serialized, forming a common type of product. These products only vary slightly from company to company.

In general, the optimal combination of technical parameters such as the specifications of the main water pipe, the span of a single-span, number of towers, and ground clearance are selected considering the unique necessities of plots, water sources, soil, crops, and other basic areas or conditions. The single-lap control area of each center pivot sprinkler irrigation system is 23.3 hm^2 to 86.7 hm^2. Adjacent tower wagons are connected by flexible joints to meet the slope operation. Each tower is equipped with a motor with a driving force between 0.75 to 1.1 kW, which is the same as an electronic control synchronization system functioning to open and close the motor on a tower. When the synchronous angle of two adjacent towers is less than 19°, the tower wagon follows one by one, rotating around the center pivot, while the sprinkler mounted on the truss conducts sprinkler irrigation simultaneously. The control area of this kind of sprinkler irrigation machine is circularly shaped, as shown in Figure 4-4 and Figure 4-5.

Figure 4-4 The center pivot sprinkler irrigation system

Figure 4-5 Aerial view of the center pivot sprinkler irrigation system

4.3.2 Performances and characteristics

The practical application of a center pivot sprinkler irrigation system has the following advantages and disadvantages.

1. Advantages

① High-level automation. This kind of sprinkler irrigation machine can save more than 90% in actual manual labor compared with surface irrigation, as well as save 25% to 75% in manual labor compared with most other kinds of sprinkler irrigation machines. It can conduct automatic irrigation, spraying day and night. Eight to twelve sprinkler irrigation machines (as many as twenty units) can be controlled simultaneously from the central control room by one person. Also, it can irrigate nearly one million acres of land efficiently, reducing the usual amount of work required.

② The conservation of water. Spray water quantities can be adjusted between 5 to 100 mm, so the spraying uniformity is high (uniform coefficient up to 85%), and the water consumption is 30% to 60% less than that of furrow irrigation.

③ Increased output. Specifically, a crop of corn's output can increase from 10% to 20%. More generally, the output increase in numerous other crops is also very significant.

④ Strong adaptability. This kind of sprinkler irrigation machine can irrigate both hillside or foothill areas with a topographic slope of up to 30%. It is nearly suitable for irrigating all crops and soil.

⑤ Multi-purpose function. This kind of sprinkler irrigation machine can spray water in combination with fertilizers and pesticides. It especially has a better spray effect of nitrogen fertilizer solution.

2. Disadvantages

① The pattern for sprinkler irrigation is circular, so the four corners of a square-shaped piece of land cannot be reached by the water. 78% of square-shaped land can be accounted for by sprinkler irrigation. There are some remedial measures, but it requires a high and unideal investment.

② In order to increase the control area of the outer ring, a long-wetted radius and high-flow rate sprinkler are generally applied. As a result, the average water application rate at the distal end of the wetted radius is very large, accumulating to more than 100 mm/h, leading to a high water application rate which is much greater than the rate of soil infiltration, increasing the possibility of surface ponds and runoff.

③ Low-pressure sprinklers are sensitive to the working pressure, and thus a pressure regulator or flow rate regulator is necessary to install to increase costs.

④ An irrigation system is required to operate continuously during the period where crops require the largest amount of water. Therefore, an adequate amount of spare parts is needed to maintain the system when it to be faulty.

4.3.3 Field design

1. Layout

It is necessary to divide irrigation areas for field layout in a way that is the most reasonable for the specificities of the field. Irrigation components, such as the span and size of the truss, the end cantilever, and the spray gun should be configured properly based on the size of irrigation land plots through technical and economic analysis so that non-irrigation area in a field is minimal.

The distance between a sprinkler irrigation machine and an adjacent facility should correspond with the shortest distance to the distal end of any adjacent sprinkler irrigation machine while being more than 3 m. The distance between the distal end of a sprinkler irrigation machine and an obstacle, such as buildings and trees, will be more than 2 m.

Water needs to avoid being sprayed from the distal end of a sprinkler irrigation machine to a road (railways, highways, etc.), traffic facilities, and power lines. Powerlines are especially important, and water sprayed farther than 6.5 m away.

2. Design

The maximum irrigation quota, design irrigation quota, design irrigation cycle, required a flow rate of controlled irrigation area for a single sprinkler irrigation machine, and demand flow rate of any sprinkler irrigation machine can be calculated according to the above Eq.(4-1) to Eq.(4-5) respectively. The designated irrigation time t should be chosen from 20 to 22 h in Eq.(4-4).

① The length of the whole sprinkler irrigation machine can be calculated as:

$$L_s = \sum_{i=1}^{j} l_i n_i + L_x \tag{4-11}$$

where L_s is the length of the whole sprinkler irrigation machine in m, l_i is the ith truss span in m, n_i is the number of the ith truss, j is the type of the truss span, and L_x is the distal cantilever length in m.

② Effective length of the sprinkler irrigation machine.

The most effective length of the sprinkler irrigation machine can be calculated as:

$$L_a = L_s + 0.75R_4 \tag{4-12}$$

where L_a is the effective length of the sprinkler irrigation machine in m, L_s is the length of the entire sprinkler irrigation machine in m, and R_4 is the wetted radius of the spray gun (or sprinkler) installed at the distal end of the sprinkler irrigation machine in m.

③ Control area of the sprinkler irrigation machine.

The control area of a working location can be calculated as:

$$A = [(\beta/360) \times \pi L_a^2]/10\ 000 \tag{4-13}$$

where A is the control area of the sprinkler irrigation machine at a working location in hm^2, and β is the fan-shaped angle of the sprinkler irrigation machine functioning in the field in degrees.

④ The shortest runtime of the sprinkler irrigation machine in a circle can be calculated as:

$$t_{min} = iL_f/(30Dn\eta) \tag{4-14}$$

where t_{min} is the shortest runtime of the sprinkler irrigation machine in a circle in h, i is the total transmission speed ratio of the drive, L_f is the distance between the center most point between the central support and the distal tower wagon in m, D is the effective diameter of the matched tire in m, n is the speed of the driving motor in r/min, and η the slip coefficient of the field, generally taken as 0.92 to 0.97.

⑤ The minimum irrigation depth of the sprinkler irrigation machine in a circle can be calculated as:

$$h_{min} = 0.1Q_0 t_{min} \eta_p / A \tag{4-15}$$

where h_{min} is the minimum irrigation depth of the sprinkler irrigation machine in a circle in mm, and Q_0 is the flow rate of the sprinkler irrigation machine in m^3/h.

⑥ The set point of a percent timer can be calculated as:

$$x = h_{min}/m \tag{4-16}$$

where x is the set point of a percent timer.

4.4 Lateral move sprinkler irrigation system

4.4.1 Structure and working principle

Lateral move sprinkler irrigation systems are mainly electrically driven while sometimes supplemented by a hydraulically-driven mode. This is the same as center pivot sprinkler irrigation systems. The electric lateral move sprinkler irrigation system is composed of a driving vehicle, tower wagon, cantilever at the distal end of the truss, electronic control synchronous system, and guiding device. Most components of the lateral move sprinkler irrigation system are the same as those of the center pivot sprinkler irrigation system, and the universal rate of components for both sprinkler irrigation machines is 85%. However, the lateral move sprinkler irrigation system has a unique structure due to different water supply modes and operating conditions.

① Central frame. The central frame is the "head" part of the lateral move sprinkler irrigation system, which acts as the center pivot sprinkler irrigation system. Two tower wagons of the central frame with stand the weight of pumping units and hanging brackets and have a left and right symmetric mode separate on both sides of the channel (or one side). The tower wagon has the same structure as the standard tower wagon, but with a slightly thicker bottom beam.

② Guide system. Unlike a center pivot sprinkler irrigation system, lateral move sprinkler irrigation systems have no fixed center support. If it is free to move in any lateral way, the lateral move sprinkler irrigation system will deviate from the channel and will not absorb water. The central frame will fall into the channel and the machine will be completely damaged. Therefore, there must be a guidance control system to constrain its displacement in its lateral direction (axis direction of the irrigator). A guidance control system mainly consists of an inching switch with a touch rod, radio tracking, and locomotive traction.

③ Sprinkler. Sprinklers on a lateral move sprinkler irrigation system are almost the same type of low-pressure sprinkler with an equidistant arrangement. The aperture of the sprinkler is slightly different; the outer end has a greater aperture to spray the same quantity of water. This is due to local restrictions through the tower wagon and the frictional head loss of pipeline.

The lateral move sprinkler irrigation system travels along a straight line. It is a kind of self-propelled sprinkler irrigation machine with a rectangular spray area. The entire irrigation area is up to 98% irrigated, which is 15% to 20% higher than that of the center pivot sprinkler irrigation system. However, it has a more complicated structure and highly oriented and synchronized requirements in operation. It needs a nearly perfect electronic control system, which is complex to operate and maintain.

A lateral move sprinkler irrigation system generally takes water from an open channel (if the source is well water, secondary water lift is needed, as several wells converge in the channel). The open channel is located in the center (or on one side) of an irrigation area. The specific number of frame units of the sprinkler irrigation machine composes a straight line set with a control center in the center (or one side) of the area, known as the central frame. The irrigation system's axis is perpendicular to the centerline of the channel when operating, so the moving trail of the tower wagon wheel of the sprinkler irrigation machine is parallel to the centerline of the channel. The central frame runs across the channel (or on the side of the channel). The diesel engine, water pump, generator, and control and guidance equipment are mounted on the hanging bracket of the tower wagon of the central frame and hanged over the channel's surface. Several standard frame units are joined on both sides of the central frame. The standard frame unit is composed of a pipeline belly frame, tower wagon, drive, control, spraying, and other components. The lateral move sprinkler irrigation system is shown in Figure 4-6.

Figure 4-6 Lateral move sprinkler irrigation system

4.4.2 Performances and characteristics

The lateral move sprinkler irrigation system has the following advantages and disadvantages in its practical application.

1. Advantages

① It can irrigate rectangular plots without missing the corners. The land utilization can be as high as 98%, while the land utilization rate of the center pivot sprinkler irrigation system united with a long-wetted radius sprinkler at the distal end of the sprinkler irrigation machine is only 87%.

② In line with traditional farming practices and plowing directions, the wheel rut has a slight effect on agricultural machinery (especially the harvester), so there is no need to smoothen the rutting before sowing. However, the circular wheel rut of the center pivot sprinkler irrigation system is detrimental to the operation of a mower, so the pasture needs to be cut 2 to 3 times a week.

③ It sprays uniformly along the direction of the irrigation area, and the wind has little effect on the quality of irrigation.

④ It is suitable for low-pressure spray, has a small head loss in pipelines, and is beneficial in reducing energy consumption. The use of a low-pressure sprinkler is also conducive to both fertilizers and pesticides.

⑤ This kind of sprinkler irrigation machine has a simple structure that doesn't have very many complex requirements. The sprinkler irrigation pipeline can be composed of pipes with different diameters, the sprinklers are arranged equidistantly on the same type, and the electric control equipment of each tower wagon is the same. Additionally, the assembly, maintenance, and repair are convenient and easy.

⑥ This kind of sprinkler irrigation machine has a high degree of automation. The length and the control area of the sprinkler irrigation machine are adjustable, which is larger than the control area of the center pivot sprinkler irrigation system with the same length. These characteristics reduce the unit area investment and energy consumption indicators.

⑦ The lateral move sprinkler irrigation system has a large adjustment range of the operating speed and can meet a variety of crop growth requirements and different water requirements.

⑧ It can be used for fertilizers, insecticides, fungicides, herbicides, plant growth regulators, and can be in partial replacement of intertillage, plant protection, and other agricultural machinery. It is a truly multi-purpose machine.

2. Disadvantages

The disadvantage of this sprinkler irrigation machine is mainly its weak ability to adapt to embankment. It requires a relatively flat ground and a guidance system to hinder traffic, so this will increase the financial costs. For example, in the use of an open channel to supply water, it will be necessary to set up pollution control equipment.

The range in the application of the lateral move sprinkler irrigation system is the same as that of the center pivot sprinkler irrigation system, but the capacity of the lateral move sprinkler irrigation system to adapt to the terrain slope is lower, while the adaptability to different soil and wind speed and direction is stronger.

4.4.3 Field design

1. Layout

For alateral move sprinkler irrigation system, it is necessary to divide the irrigation areas into several rectangular plots in field layout based on the type of irrigation and irrigation capacity. Irrigator components such as sprinkler span and the number of trusses, end cantilevers, and spray guns should be configured properly based on the size of irrigation land plots, which can be determined through technical and economic analysis.

The distance between the lateral move sprinkler irrigation system and adjacent facilities should correspond with the shortest distance between any two adjacent sprinkler irrigation machines while being more than 5 m. The distance between the distal end of the sprinkler irrigation machine and an obstacle, such as buildings and trees, will be more than 2 m. Water needs to avoid being sprayed from the distal end of the sprinkler irrigation machine to a road (railways, highways, etc.), traffic facilities, and power lines. Powerlines are especially important, and water sprayed farther than 6.5 m away.

When the slope of the parcel in driving direction is greater than or equal to 1‰, the watering method of utilizing a dragging hose should be adopted or the relevant measures should be taken to meet the water requirements of the pump. When the slope of the parcel in the walking direction is less than 1‰, the utilization of a channel water supply should be adopted.

When the demand flow rate of the lateral move sprinkler irrigation system is less than 130 m^3/h, the water supply of pipes can be selected for the irrigation land. When the demand flow rate is more than 130 m^3/h, the water supply of canal water can be used. Supposing the length of the whole sprinkler irrigation machine is less than 350 m, the one-sided water supply method can be used for irrigation land. Finally, if the length of the entire sprinkler irrigation machine is more than 350 m, the double-side water supply method should be preferred in the irrigation land.

If a dragging hose is used for the water supply by the sprinkler irrigation machine, a passage for the trolley to travel while dragging the hose should be created on one side of the water supply pipe. The arrangement of the water supply pipe's axis and the hydrant should be parallel to the direction of the path of the sprinkler irrigation machine. The length of the water supply pipe should meet the operational requirements of the control plots of the irrigation. Hydrants on the water supply pipes should be arranged in equal spacing with an interval ranging from 50 to 100 m.

If the piped water supply method is adopted by the sprinkler irrigation machine, a passage for the trolley to travel while dragging the hose should be made on one or both sides of the water supply pipe. The water supply channel should be central to the canal and parallel with the traveling direction of the sprinkler irrigation machine. The channel cross-section should be conducted with anti-seepage treatment. The channel length should meet the operation requirements of the control plots of irrigation. Channel width and depth should meet the water requirements of the pump on the drive trolley.

2. Design

The maximum irrigation quota, design irrigation quota, design irrigation cycle, required a flow rate of the controlled irrigation area for a single sprinkler irrigation machine, and demand flow rate of the sprinkler irrigation machine can be calculated according to the above Eq.(4-1) to Eq.(4-5) respectively. The designated irrigation time t should be between 20 to 22 h, represented in Eq.(4-4).

① The most effective length of the sprinkler irrigation machine can be calculated as:

$$L_a = L_s + 0.75(R_5 + R_6) \tag{4-17}$$

where L_a is the effective length of the sprinkler irrigation machine in m, L_s is the distance between the main drive trolley and the sprinkler (spray gun) at the distal end of the water pipe of a truss for a one-sided lateral move sprinkler irrigation system in m. it is also the distance between sprinklers (spray guns) at the two distal ends of the truss hose on both sides of the sprinkler irrigation machine for a bilateral lateral move sprinkler irrigation system in m, and R_5 and R_6 are the wetted radius of both sides of the terminal sprinkler (spray gun) in m.

② The designated irrigation time along the irrigation block length for one watering period can be calculated as follows:

$$t_1 = 0.001 m L_a L_b / (Q_0 \eta_p) \tag{4-18}$$

where t_1 is the designated irrigation time along the irrigation block length for one watering period in h, L_b is the length of the irrigation plots in m, and Q_0 is the demand flow rate of the sprinkler irrigation machine in m^3/h.

③ The minimum irrigation depth along the irrigation block length for one watering period can be calculated with Eq.(4-19) and Eq.(4-20):

$$h_{min}=1\,000Q_0 t_{min}\eta_p/(L_a L_b) \tag{4-19}$$

$$t_{min}=iL_b/(60\times\pi nD\eta) \tag{4-20}$$

where h_{min} is the minimum irrigation depth along the irrigation block length for one watering period in mm, and t_{min} is the smallest allotted time along the irrigation block length for one watering period in h.

④ The set point of a percent timer is calculated as follows:

$$x=h_{min}/m \tag{4-21}$$

where x is the set point of a percent timer.

4.5 Side-roll wheel sprinkler irrigation system

4.5.1 Structure and working principle

The structure of a side-roll wheel sprinkler irrigation system is shown in Figure 4-7[14]. It is mainly composed of a motor vehicle, water pipe, wheels, automatic drain valve, sprinkler, sprinkler counterweight, and terminal plug.

The motor vehicle and water pipes are connected through a host joint, and the water pipes are fastened to each other through triangular flanges. The wheels are mounted on the welded steel angle at the end of each water pipe, so the water pipes work as axles. The water pipe is fixed with a sprinkler counterweight, which is mounted on, and an automatic drain valve. The terminal plug is installed at the distal end of the water pipe, while the other end is connected to the water diversion hose and hydrant.

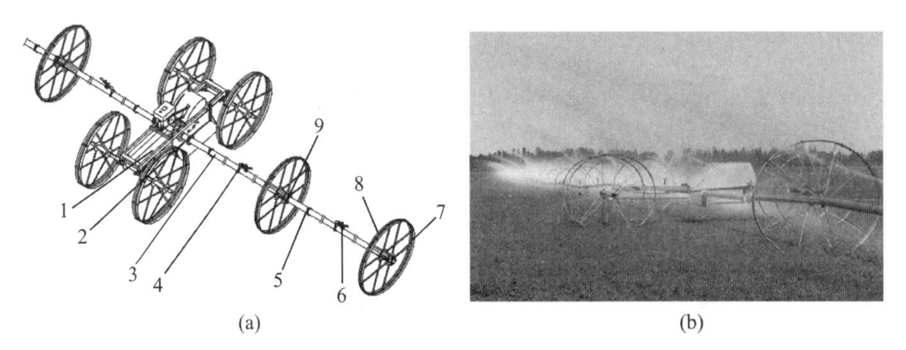

1—Motor vehicle; 2—Host joint; 3—Water pipe; 4—Spray head counterweight; 5—Automatic drain valve; 6—Spray head; 7—End plug; 8—Wheels in branch pipe; 9—Triangular flange joint

Figure 4-7 Side-roll wheel sprinkler irrigation system

The operation of a side-roll wheel sprinkler irrigation system utilizes sprinklers working intermittently. The side-roll wheel sprinkler irrigation system can function when the irrigation volume reaches its specific requirement; the water control valve of the water pipe is closed, the connection between the initial end of the water pipe and the water main pipe is disconnected, and the remaining water is discharged by the automatic drain valve. When the remaining water in the water pipe is less than 20%, the water is considered to have been drained out. The operating system works to drive the whole machine to the next sprinkling irrigation area, where it connects the initial end of the water pipe with the main water pipe, opens the water control valve, and again starts the fixed sprinkler irrigation at the new location. This cycle will repeat until irrigation has been completed.

4.5.2 Performances and characteristics

The control area of the side-roll wheel sprinkler irrigation system is influenced by many factors, such as crop water demand, pipe size, soil type, and so on. In general, a side-roll wheel sprinkler irrigation system can irrigate from 10 to 50 hm^2 of the land with moving pipes up to 600 to 800 m in length.

The designated water application rate of a side-roll wheel sprinkler irrigation system should not exceed the allowable water application rate of the specific type of soil to be irrigated. The water supply flow should be larger than the total designated flow rate of the irrigation system, and this is the same for the total water supply. When the water supply flow rate is less than the total design flow rate of irrigation systems, and the total amount of water supply is greater than the total amount of design water for irrigation systems, storage facilities are needed to conserve the extra water.

The amount of water required for a side-roll wheel sprinkler irrigation system depends on the area of irrigation and the water demand of the crop. In general, a machine set with a pipe length of 400 m and an irrigation area of 15 hm^2 will have a flow rate of 36 to 50 m^3/h, while a machine set with a pipe length of 600 m and an irrigation area of 30 hm^2 will have a design flow rate of 72 to 108 m^3/h.

The working pressure of a side-roll wheel sprinkler irrigation system is generally between 0.3~0.5 MPa. The actual working pressure of any sprinkler should not be lower than 90% of the designated working pressure of the sprinkler. The working pressure difference between any two sprinklers on the same side-roll wheel sprinkler irrigation system should be within 20% of the designated working pressure of the sprinkler.

The amount of irrigation at each operation position is controlled by the irrigation time, so that different amounts of irrigation can be most accurately applied to seedlings and mature crops. For example, a machine set with a pipe length of 400 m and the flow rate of 75 m^3/h will provide a watering depth of 63.5 mm after watering for 6 hours. This will meet the water demands of most crops.

4.5.3 Field design

1. Layout

For the field layout of a side-roll wheel sprinkler irrigation system, it is recommended that a farmer divide an irrigation area into several rectangular plots based on the type and the irrigation capacity of the sprinkler irrigation machine. The sprinkler irrigation machine type, effective length, and irrigation control area can be reasonably determined through technical and economic analysis. The length of the whole machine should not exceed the maximum length recommended by the supplier of the system. The driving car should be set in the middle of the entire machine's length.

The most advantageous water supply for the side-roll wheel sprinkler irrigation system is through pipes. The laying direction of the water supply pipeline should be perpendicular to the axle branch pipelines. Two adjacent sprinkler irrigation machines can share water supply pipelines and hydrants. The distance between hydrants should be an integral multiple of the effective outer perimeter of the roller and should not exceed 1.3 times the range of the matched sprinkler under the designed working pressure.

2. Design

The maximum irrigation quota, the designated irrigation quota, the designated irrigation cycle, the required flow rate of a controlled irrigation area for the single sprinkler irrigation machine, and the demand flow rate of the sprinkler irrigation machine can be calculated utilizing Eq.(4-1) to Eq.(4-5) respectively.

① The distance between two adjacent operating positions of the sprinkler irrigation machine can be calculated using Eq.(4-22). The results should be adjusted to an integer that are multiples of the effective outer perimeter of the roller.

$$b_1=k_bR_1 \tag{4-22}$$

where b_1 is the distance between two adjacent operation positions of the sprinkler irrigation machine in m, k_b is the ratio of the distance between the positioning spray and the sprinkler, ranging from 0.8 to 1.3 according to the wind speed, and R_1 is the sprinkler wetted radius in m.

② The positioning spray length of the sprinkler irrigation machine can be determined as follows:

$$L_a=L_s+2kR_1 \tag{4-23}$$

where L_a is the positioning spray length of the sprinkler irrigation machine in m, L_s is the length of the whole sprinkler irrigation machine in m, k is the reduction factor of the wetted radius, which is related to the wind speed and can be found in Table 4-2.

③ Water application rate can be calculated according to Eq.(4-24) to Eq.(4-25):

$$\rho=K_wC_\rho\rho_s \tag{4-24}$$

$$K_w=1.08v^{0.194} \tag{4-25}$$

$$K_w=1.12v^{0.302} \tag{4-26}$$

$$C_\rho=\pi/\{\pi-\pi/90\times\arccos[a/(2R_1)]+a/R_1\times\sqrt{1-(a/(2R_1)^2}\} \tag{4-27}$$

$$\rho_s=1\ 000q/(\pi R_1^2) \tag{4-28}$$

where ρ is the water application rate in mm/h, K_w is the wind coefficient, which is specifically calculated using Eq. (4-25) when the direction of the axle branch pipe is perpendicular to the wind direction or calculated using Eq.(4-26) when the direction of the axle branch pipe is parallel to wind direction, C_ρ is the combination coefficient, v is the wind speed in m/s, a is the measured space between sprinklers in m, and ρ_s is the designated water application rate for a single sprinkler spraying circularly without wind in mm/h.

④ Irrigation time of the sprinkler irrigation machine at a working position can be calculated as follows:

$$t_1 = 0.001 m L_a b_1 / (Q_0 \eta_p) \tag{4-29}$$

where t_1 is the irrigation time of the sprinkler irrigation machine at a working position in h, and Q_0 is the demand flow rate of the sprinkler irrigation machine in m^3/h.

References

[1] Yan H J. Study on water distribution uniformity of center pivot and lateral move irrigation systems based on variable rate technology[D]. Beijing: China Agricultural University, 2004.

[2] Wang Y H. Key components design and experimental research of roll wheel line move irrigator[D]. Heilongjiang: Northeast Agricultural University, 2016.

[3] Tang L D. Optimization design and analysis of inner flow characteristics for the special water turbine used in hose reel irrigator[D]. Zhenjiang: Jiangsu University, 2017.

[4] Tang Y. Research and development of intelligent agricultural irrigation equipment in automatic coiling type with low energy[R/OL].[2016 - 03 - 01].http://www.jsstrs.cn/xiangxiBG.aspx? id=2281.

[5] Tang L D. Numerical simulation on characteristics of secondary flow resistance for a hose reel irrigator[D]. Zhenjiang: Jiangsu University, 2013.

[6] Gu Z, Tang Y, Tang L D. Calculation and analysis on load of hose reel irrigators [J]. Journal of Agricultural Mechanization Research, 2015, 8: 70 - 73.

[7] Zhang Y S, Zhu D L, Zhang L. Study on translocating speed and water distribution uniformity of lightweight lateral move irrigation system[J]. Journal of Drainage and Irrigation Machinery Engineering, 2014, 32(7): 625 - 630.

[8] Li Y B, Xie C B, Zhang G H. Research and development of self-driven spray and drip irrigation equipment with multifunction [J]. Water Saving Irrigation, 2014, 10: 89 - 91.

[9] Fan Y S, Huang X Q, Wu F. Water distribution characteristics and experiments of spray irrigation and hose irrigation dual-use unit [J]. Transactions of the Chinese Society for Agricultural Machinery, 2009, 40(11): 74 - 77.

[10] Xu Y F, Francis T. Characteristic analysis of automatic hose reel irrigator [J]. Water Saving Irrigation, 1979, 3: 39 - 45.

[11] Tang L D, Yuan S Q, Tang Y. Analysis on research progress and development trend of hose reel irrigator[J]. Transactions of the Chinese Society for Agricultural Machinery, 2018, 49(10): 8 - 22.

[12] Wang Y H, Sun P L, Sun W F. Key components design and experiment of roll wheel line move sprinkling irrigation machine[J]. Journal of Drainage and Irrigation Machinery Engineering, 2015,33(10): 915 - 920.
[13] SL 280—2019. Technical specification for application of large and medium-scale sprinkler irrigation system[S].
[14] Irrigation engineering design manual editors. Irrigation engineering design manual[M]. Beijing: Water Resources and Electric Power Press,1989.

CHAPTER 5
Micro-irrigation System and Equipment

5. 1 Introduction

5.1.1 Types of micro-irrigation

Micro-irrigation is a local irrigation technique that distributes pressurized water to a field using a system assembled by micro-irrigation equipment. This equipment wets the soil near the root zone of the crop with a lightly pressured flow using a douche[1-3]. Micro-irrigation can be classified by different types of equipment (mainly emitters) and different forms of outflow, there are drip irrigation, microspray irrigation, small pipe irrigation, and infiltration irrigation[4-8].

1. Drip irrigation

Drip irrigation uses emitter installed at the distal end of pipelines or uses drip irrigation hose integrated with laterals to wet soil with drop pressured water with a fine flow in the situation where the emitter flow rate is too large. Usually, the laterals and emitters are either placed on the ground or buried 30 to 40 cm below the ground. The former is called surface drip irrigation, and the latter is called subsurface drip irrigation. The drip irrigation emitter has a flow rate of 2 to 12 L/h.

2. Microspray irrigation

Microspray irrigation uses minisprinklers installed on laterals directly or connected with laterals to wet soil with spray pattern pressured water. There are two types of minisprinklers: fixed and rotating. Fixed minisprinklers have a small spray wetted radius and small water droplets while rotating minisprinklers have a larger spray wetted radius and larger water droplets, so the installation spacing is also large. The flow rate of a minisprinkler is usually between 20 to 250 L/h.

3. Small pipe irrigation

Small pipe irrigation uses an emitter comprised of small plastic tubes and laterals to wet the soil near the crop with a trickle (jet) shape. Small pipe irrigation has a flow rate between 20 to 250 L/h.

4. Infiltration irrigation

Infiltration irrigation uses special porous pipes buried 30 to 40 cm below the soil's surface. The pressured water passes through the pores of the pipe to moisten the soil in the form of seepage. Infiltration irrigation reduces the evaporation that may occur on the soil's surface, therefore it has the least water consumption. The flow rate of infiltration irrigation porous pipes ranges from 2 to 3 L/(h · m).

5.1.2 Advantages and disadvantages of micro-irrigation

1. Advantages

Micro-irrigation wets the soil very close to where plants are rooted and maintains a lower water pressure to meet the needs of crop growth regularly. Micro-irrigation also has the following advantages[9-13].

(1) Conserves water and reduces labor and energy costs

Micro-irrigation irrigates in a timely and appropriate fashion which best fits the requirement of crop storage, while only wetting the soil near the most significant area of the individual crops or plants so that water loss is reduced significantly. Micro-irrigation is the most easily operated water supply for a pipe network, while also the most convenient as it has a high labor efficiency, and convenient automatic control. This convenience ultimately saves a lot of labor. Simultaneously, micro-irrigation is local irrigation; most of the land surface is dry, which reduces the growth of weeds, and the associated labor and herbicide costs of weeding. Fertilizer and pharmaceuticals are applied to the soil near the root directly with irrigation water through the irrigation systems, so there is no need for manual operation, improving the efficiency and utilization rate of the application of fertilizer. The working pressure of a micro-irrigation emitter is between 50 to 150 kPa, which is significantly less than the working pressure of a typical sprinkler irrigation system. Compared to surface irrigation, micro-irrigation saves more water and reduces the energy consumption and costs associated with it.

(2) Even irrigation application

A micro-irrigation system can effectively control the outlet water flow of each emitter, so the irrigation uniformity is high, generally reaching 80% to 90% uniformity.

(3) Increased production

Since micro-irrigation systems supply water and fertilizer to the root zone of a crop in a timely and appropriate manner, the condition of the water, heat, gas, and soil nutrients are superior, so a large and stable production can be achieved to improve the quality of crop production.

(4) Possible irrigation with saline water

In micro-irrigation, the soil water content in the active layer of a crop's roots is always maintained in the most favorable condition for crop growth, so any soil salinity can be diluted. Micro-irrigation may be the best solution when the water has high salinity.

(5) Strong adaptability to soil and topography

The douche should be selected depending on the most influential characteristics of an irrigated area's soil so the corresponding emitter can be adjusted without runoff and deep leakage. Micro-irrigation uses pressured pipes to delivered water to every tree near the roots of crops, which can work effectively in any complex terrain conditions and can even be used on areas with a steep topography or riprap beach trees.

2. Disadvantages

The investment in micro-irrigation system is generally much higher than that of surface irrigation. Additionally, the emitter has a small outlet, so it is easily blocked by mineral organic matter in water. If this were to occur, the water distribution uniformity of system would be reduced, leading to the system not working properly, even possibly having to be discarded all together.

5.2 Composition and classification of micro-irrigation system

5.2.1 Composition of micro-irrigation system

A micro-irrigation system consists of a water source, headwork of pivot, transmission and distribution pipe network, emitter, flow rate, and pressure control components, and measuring instrument, as shown in Figure 5-1.

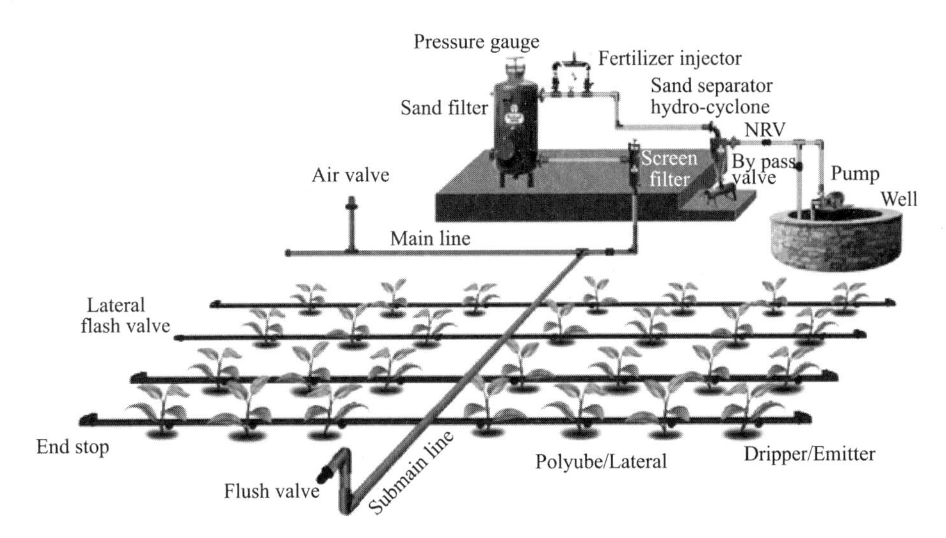

Figure 5-1 Schematic diagram of micro-irrigation system

1. Water source

Rivers, channels, lakes, reservoirs, wells, springs, etc. can be used as irrigation water sources, but their water quality should meet the requirements of micro-irrigation.

2. Headwork of pivot

The headwork of pivot is made up of a pump, power machine, fertilizer, and chemical injection equipment, filtration equipment, control valve, exhaust valve, and pressure rate and flow measuring instruments. Its function is to pump water from a water source and to treat pressurized irrigation water to the entire system.

The most commonly used pumps for micro-irrigation include submersible pumps, deep well pumps, and centrifugal pumps. The power machine can be a diesel engine or an electric motor. A pump is not necessary for an area where there is enough natural water head.

Reservoirs and sedimentation tanks are often needed when the water supply needs to be regulated or the sediment concentration is large. The sedimentation tank is used to remove the large solid particles in the irrigation water. In order to avoid producing algae in the sedimentation tank, the tank or reservoir should be covered as much as possible.

The function of the filter equipment is to filter the solid particles of the irrigation water to prohibit dirt from entering the system, which would result in system congestion. The filter equipment should be installed before the water supply and distribution pipeline.

Fertilizer and chemical injection equipment are used to direct fertilizers, herbicides, pesticides, etc. into the micro-irrigation system; the injection equipment should be installed before the filtration equipment.

Flow rate and pressure measuring instruments are used to measure the flow rate and pressure in pipelines, including flowmeters and pressure gauges. The flowmeter is used to measure the total amount of water flowing through the pipeline. It can be installed on the head or on any trunk or branch pipe depending on what is necessary. If it is installed in the head unit, it must be placed before the fertilizer and water equipment to prevent them from corroding. The pressure gauge is used to measure pipelines' internal pressures. A pressure gauge is installed before and after the filter and sealed fertilizing device so that the pressure difference can be accurately measured. The size of the fertilizer filter and how it should be cleaned can be determined by the amount of pressure difference.

The controller is used to control the irrigation system automatically; the general controller has the function of timing and programming. The solenoid valve (or hydraulic valve) is operated according to the instruction manual provided by the system's manufacturer, so the system can be controlled.

The valve is used to control and adjust the pressure and flow rate of the micro-irrigation system directly. A valve is usually installed in the parts most needing control. Generally, gate valves, check valves, air valves, hydraulic valves, and solenoid valves are used.

3. Transmission and distribution pipe network

The function of the water distribution pipe network is to transfer the treated water to each irrigation unit and emitter according to its specific necessities and requirements. The water distribution pipe network consists of three pipelines: run pipe, branch, and lateral. The lateral is the last leveled pipe of the micro-irrigation system that is installed, which is connected with the emitter. In the micro-irrigation system, polyethylene pipes are commonly used when the pipe diameters need to be less than or equal to 63 mm, and Polyvinyl Chloride (PVC) pipes are used for diameters more than or equal to 63 mm.

4. Emitter

The emitter is one of the most important components of the micro-irrigation equipment, which is applied to distribute water to the plant directly. Its role is to reduce the pressure and transform the water flow to drip, trickle or spray before it enters into the soil. The emitter is made up of a minisprinkler, emitter, and drip irrigation hose. Most emitters are made of plastic.

5.2.2 Classification of micro-irrigation system

Because emitters of micro-irrigation systems are different and unique, the corresponding micro-irrigation systems can be divided into four categories: the drip irrigation system, microspray irrigation system, small pipe irrigation system, and infiltration irrigation system[14-17].

Depending on whether the water distribution pipeline moves or not in the irrigation season, each kind of micro-irrigation system can be sub-divided into fixed, semi-fixed and mobile systems. All parts of the micro-irrigation system are fixed throughout the irrigation season; generally primary and branch pipes are buried in the ground. Depending on different conditions of each system, some laterals are also buried in the ground, while some are placed on the surface, and others are hung on the high brackets a few centimeters off the ground. The fixed micro-irrigation system is usually utilized for crops with a higher economic value. In the semi-fixed micro-irrigation system, the headwork of pivot, trunk, and branch pipes are fixed, while the laterals and the emitters on them are portable. Based on the design of the system, a lateral can work in different positions. All components of a mobile micro-irrigation system can be moved, and during an irrigation cycle, they are arranged in different positions throughout the area to irrigate according to plan. Semi-fixed and mobile micro-irrigation systems can improve the utilization rate of the micro-irrigation equipment and reduce the irrigation investment per unit area. They are commonly used for field crops. But due to the inconveniences of the actual operation of these systems, they are more suitable for drought-ridden and poorly conditioned areas.

5.3 Emitter

The role of the emitter is to pour pressured water through the lateral into the soil near the crop's root zone equably and steadily[18-21]. The quality of an emitter directly affects the service life and irrigation quality of the micro-irrigation system as a whole. There are many kinds of emitters, each of which has specific characteristics and working conditions.

The requirements of emitter: ① the manufacturing deviation is little, and the deviation coefficient C value of the emitter should be generally controlled below 0.07. ② The output of water is stable with the small affection of the water head. ③ It needs to have strong anti-clogging ability. ④ The structure is simple and easy to manufacture, install and clean. ⑤ It is durable and is not costly.

5.3.1 Structural parameters and hydraulic performance

Two main technical parameters of micro-irrigation emitters are the structural and hydraulic performance capacities. The structural parameters mainly refer to the size of the runner or orifice, and the hose diameter and wall thickness are also included in the drip irrigation hose. The hydraulic performance parameters mainly refer to the flow index, manufacturing deviation coefficient, working pressure, flow rate, and the sprinkler also include the wetted radius, water application rate, water distribution, and so on. Table 5-1 lists the structural and hydraulic performances of all kinds of emitters, available for reference. The C_v value is stipulated in industry-standard SL/T 67.1~3—94.

Table 5-1 Technical parameters of micro-irrigation emitter

Emitter type	Structural parameters					Hydraulic performance parameters				
	Runner or orifice diameter r/mm	Runner length h/cm	Droppers or orifices spacing/cm	Pipe diameter/mm	Wall thickness of pipe/mm	Working pressure/kPa	Outflow/(L/h) or [L/(h · m)]	X	C_v	Wetted radius/m
Dropper	0.5~1.2	30~50				50~100	1.5~1.2	0.5~1.0	<0.07	0.5~4.0
Drip tape	0.5~0.9	30~50	30~100	10~16	0.2~1.0	50~100	1.5~3.0	0.5~1.0	<0.07	
Micro sprinkler	0.6~2.0					70~200	20~250	0.5	<0.07	
Water generator	2.0~4.0					40~100	80~250	0.5~0.7	<0.07	
Infiltrating irrigation pipe(belt)				10~20	0.9~1.3	40~100	2~4	0.5	<0.07	
Pressure compensated type								0~0.5	<0.15	

Note: the outflow of the infiltration irrigation pipe (belt) is calculated in L/(h · m), other flow rate ares in L/h. All emitters have a specific type of pressure compensation, but the parameters apply to all, generally $X<0.3$ for full compensation, otherwise partial compensation.

1. Relationship between flow rate and working pressure

The relationship between flow rate and working pressure of a micro-irrigation emitter is given as:

$$q=kh^{x} \tag{5-1}$$

where q is the emitter flow rate, h is the working head, k is the flow rate coefficient, and x is the flow index.

In Formula (5-1), the flow index x reflects the sensitivity of the emitter flow rate to the change of working pressure. When the dripper has a laminar flow, the flow index x is equal to 1.0. This means the flow rate is proportional to the water head. When the dripper flow is fully turbulent, the flow index x is equal to 0.5, while the flow index x of the full pressure compensation emitter is equal to zero (0), meaning the outer water flow is not affected by the pressure changes. The flow index of other kinds of emitters varies from 0 to 1.

2. Manufacturing deviation coefficient

The flow rate of the emitter is proportional to 2.5 to 4 times the power of the flow diameter, and any small deviation caused by manufacturer will cause a larger flow deviation. In the manufacture of irrigation machines, deviations inevitably arise due to the normal occurrences of the manufacturing process and the shrinkage deformation of some material used. In practice, the manufacturing deviation coefficient is generally used to measure the manufacturing accuracy of the product. Its calculation is shown below:

$$C_v = \frac{S}{q} \tag{5-2}$$

$$S = \sqrt{\frac{1}{n-1} \sum_{i=1}^{i=n} (q_i - \overline{q})^2} \tag{5-3}$$

$$\overline{q} = \frac{\sum_{i=1}^{i=n} (q_i - \overline{q})^2}{n} \tag{5-4}$$

where C_v is the deviation coefficient of the emitter, S is the standard deviation of the flow rate, q_i is the flow rate of each measured emitter in L/h, and n is the number of measured emitters.

5.3.2 Common emitters

1. Emitter and drip irrigation hose

(1) Basic types

① Side-fitted emitter

Figure 5-2 shows an emitter inserted in a lateral, referred to as a side-fitted emitter. The structural feature of this emitter is a very short run, and the working pressure is mainly consumed in the local loss of the inlet and outlet. It belongs to the orifice emitters which have two structures: non-pressure compensation and pressure compensation. It consists of the emitter-base and the cover. A small piece of a circular elastic diaphragm (usually a silicon film) is placed at the top of the cover to make the emitter able to determine the pressure compensation.

Figure 5-2 A mounted side-fitted emitter

The role of the elastic diaphragm is to separate the emitter channel into two areas. The flow cross-section of the lower flow passage of the elastic diaphragm is much larger than that of the inlet side of the upper channel inlet. The water flows from the lateral into the emitter, through the lower runner region and then into the upper flow channel, and finally out from the outlet of the emitter.

When the working pressure of the emitter increases, the flow velocity in a lower region of the elastic is much smaller than that in an upper region, the pressure is greater than in the upper side, forcing the elastic deformation of the diaphragm. It becomes convex to the top inlet, which reduces the cross-section and makes the flow stable. On the other hand, when the working pressure of the emitter decreases, so does the elastic deformation of the diaphragm; the inlet cross-section of the top cover runner increases, and as a result, the flow velocity is stable and unchanging. The main factors affecting the compensation performance of the emitter are the cross-section size of the runner and the thickness and elasticity of the diaphragm. Therefore, it is vital to choose a high-quality elastic diaphragm.

The side-fitted emitter is flexible in the field, and the emitter can be installed based on any effective position that a specific plant requires. For widely spaced and tall plants, such as trees, using this kind of emitter is advantageous and especially suitable for a drip irrigation system with undulating topography. Unfortunately, though, manual installation of the system is time-consuming and can easily fall off. When conditions permit, an emitter can be installed on a lateral following the required spacing and then installed in the field to improve the efficiency.

② Arrow-shaped emitter

An arrow-shaped emitter is referred to as an arrow emitter. Figure 5-3 shows the structure of several kinds of arrow emitters. The upper section of the emitter is an emitter with a runner, and the lower section is a diamond insert with a pointed edge. When in use, the upper section is inserted with PE plastic microtubules and connected with the water supply pipe, while the lower section is inserted into the soil to directly dampen it with a small flow rate. This kind of emitter is simple in structure, flexible in use and low in price. It is very suitable for plants assembled in wide rows, especially potted plants. Figure 5-4 shows the shape and assembly of an arrow emitter. Figure 5-5 shows the case where the arrow emitters are used for potted flowers.

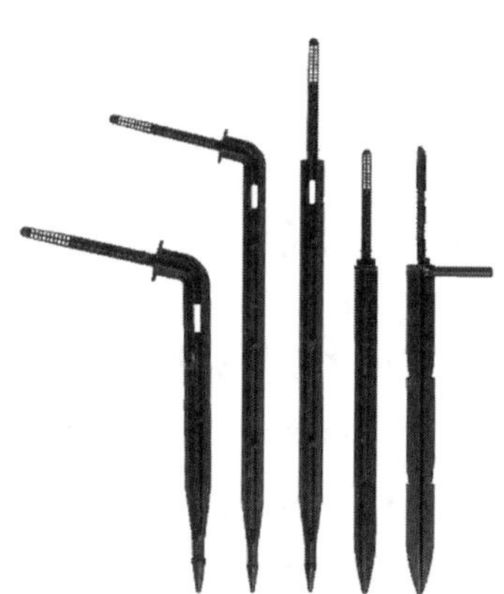

Figure 5-3 The structure of several kinds of arrow emitters

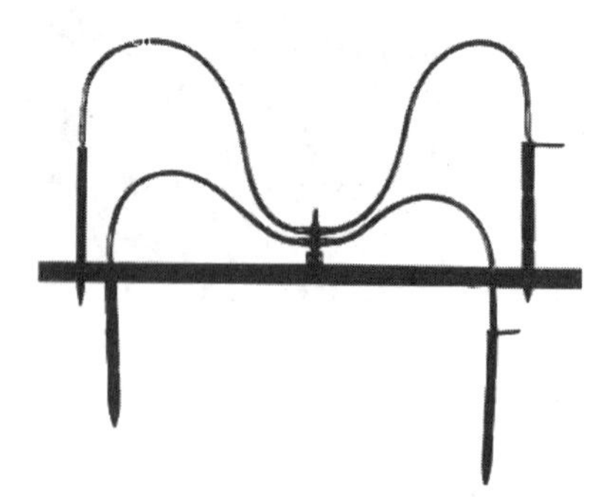

Figure 5-4 The shape and assembly of an arrow emitter

Figure 5-5 Arrow emitters being used for potted flowers

③ In-line emitter

Figure 5-6 shows an in-line emitter with two ends connected to laterals. This kind of emitter has a long flow channel, meaning it is a type of emitter with a long path. It has both dripping and water traverse capabilities. This kind of emitter can be installed following any sizeable plant row spacing in a field, so it is suitable for widely spaced plants. Unfortunately, it is easily disengaged when moving.

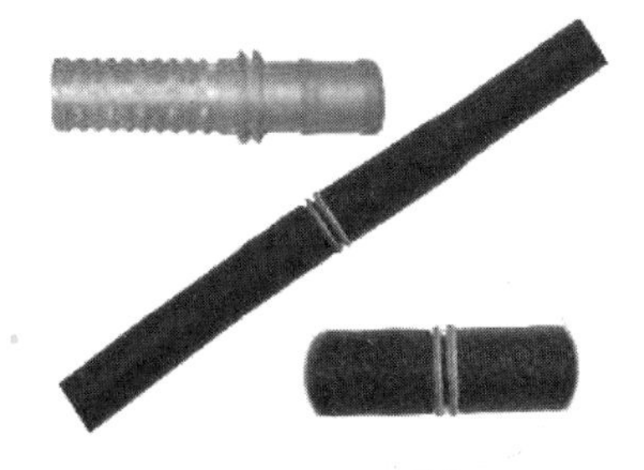

Figure 5-6 In-line emitter

④ Inner inlay emitter and drip irrigation hose

The emitter is embedded in alateral at a certain distance and integrated into it. Because of the different flow channel structures, these emitters also have non-compensation and compensation types. This kind of emitter overcomes the disadvantages of drip irrigation hoses, which are normally susceptible to falling off when moving, while it is also capable of improving field installation efficiency.

Based on the specific structure of the emitter and the imply method, the inner inlay emitter and the corresponding drip irrigation hose can be categorized into two different types.

The sheet emitter and surface-mount drip irrigation hose. The shape of the sheet emitter and the surface-mount drip irrigation hose are shown in Figure 5-7.

The tubular emitter and drip irrigation hose. It is around emitter with a labyrinth channel around the outer wall, embedded in the lateral with a certain distance. The structure of this emitter is shown in Figure 5-8.

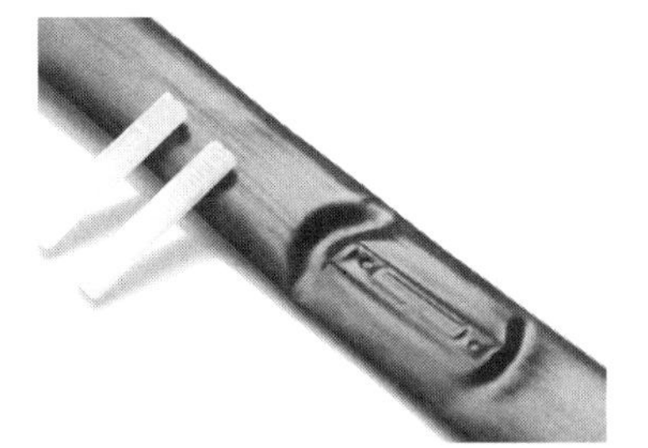

Figure 5-7 Sheet emitter and surface-mount drip irrigation hose

Figure 5-8 Tubular emitter and drip irrigation hose

⑤ Edge-pressing emitter and drip irrigation hose

A drip irrigation hose is extruded from the long labyrinth channel emitter at one side of the lateral with some intervals. This drip irrigation hose with an edge drop is formed using compression molding technology. It has a simple manufacturing process while not being costly, and it is presently the "disposable" drip irrigation belt widely used for field crops, referred to as the single-wing. Figure 5-9 shows the structure and shape of the drip irrigation hose.

Figure 5-9 Drip irrigation hose with edge drop

(2) Main technical indexes and performance parameters

The main technical specifications for emitters are shown in Table 5-2, derived from GB/T 17187—2009 and SL/T 67.1-94.

Table 5-2 The main technical indicators of the emitters

<table>
<tr><th colspan="2">Project</th><th>Qualified indicators</th><th colspan="2">Test conditions</th></tr>
<tr><td colspan="2">Rated flow rate or nominal flow rate/ (L/h)</td><td></td><td rowspan="6">The test water temperature is (23+2)℃, and filtered by 160 to 200 mesh screen or filter recommended by the manufacturer</td><td rowspan="3">The test pressure is 0.5 to 1.5 times the rated working pressure</td></tr>
<tr><td colspan="2">Rated working pressure or working pressure range/kPa</td><td></td></tr>
<tr><td rowspan="2">Hydraulic performance</td><td>Flow index x</td><td></td></tr>
<tr><td>Flow rate coefficient k</td><td></td><td rowspan="3">Mean value of the maximum and minimum working pressure</td></tr>
<tr><td rowspan="2">Structural performance</td><td>Manufacturing the deviation coefficient C_v</td><td>$\geqslant \mid \pm 0.07 \mid$</td></tr>
<tr><td>Flow path and orifice diameter / mm</td><td></td></tr>
<tr><td colspan="2">Resistance to pull</td><td>When the axial connection is applied, the axial tensile force at both ends of the specimen is applied to the maximum in 30 s, without damage or detachment; the pulling force 30 N on the pipe is not damaged and not broken in the 30 s</td><td>At normal temperature, connect with PE plastic pipe referencing the connection mode provided by the manufacturer</td><td></td></tr>
</table>

Depending on the actual needs of users, the technical performance parameters of the emitter supplied by the manufacturer shall include the rated working pressure, rated flow rate, flow index x, flow rate coefficient k (or the relationship between the flow rate and pressure formula) and manufacturing deviation coefficient C_v, and applicable working pressure range. For drip irrigation hoses, the hose diameter (mm), normal service pressure, and maximum pressure capacity are also idled. However, many manufacturers have not been able to provide all the technical parameters required, which will undoubtedly affect the user's proper application of the emitters. Table 5-3 is the technical parameters for drip irrigation emitters for reference.

Table 5-3 The technical parameters of partial emitters

Name/Model	Structural features and installation methods	Nominal flow rate or rated flow rate/(L/h)	Flow rate coefficient k/(Pressure units kPa)	Flow index x	Manufacturing deviation coefficient C_v	Working pressure range/kPa	Data sources
Pressure compensation emitter	Lateral inserting	6	4.504	0.051 9		50～450	Gan Su province Da Yu 2006 year sample
Pressure compensating dripper	Lateral inserting	8	5.408	0.073 9		50～450	
Pressure compensation emitter	Lateral inserting	10	5.926	0.102 5		50～450	
Non-pressure compensation emitter	Inlay patch	0.8	0.079 6	0.502 4		30～120	
		1.38	0.139 8	0.480 5			
		1.8	0.147	0.546 1			
		2.4	0.247	0.494			
		3	0.305 8	0.477 3			
DHD-2 pressure compensation emitter	Lateral inserting	2	1.92	0.04	0.048	40～280	Jie Yang Da HuaSend sample data
DHD-4 pressure compensation emitter	Lateral inserting	4	1.61	0.18	0.042	40～280	
DHD-6 pressure compensation emitter	Lateral inserting	6	3.45	0.12	0.045	40～280	
Pressure compensation emitter	Lateral inserting	4	3.025	0.05	0.029	100～400	Fu Zhou Wei Yu Run send sample data

2. Tubule outflow device

The tubule outflow device is the key component of a tubule outflow irrigation system. It utilizes a combination between a stabilizer and a tubule, transferring the pressured water flow from the water distribution network, easing the water into a trickle, and implementing the irrigation device for the outflow irrigation of the tubule.

(1) The composition of a tubule outflow device

The tubule outflow device consists of two parts: a flow regulator and a $\phi 4$ PE tubule, as shown in Figure 5-10.

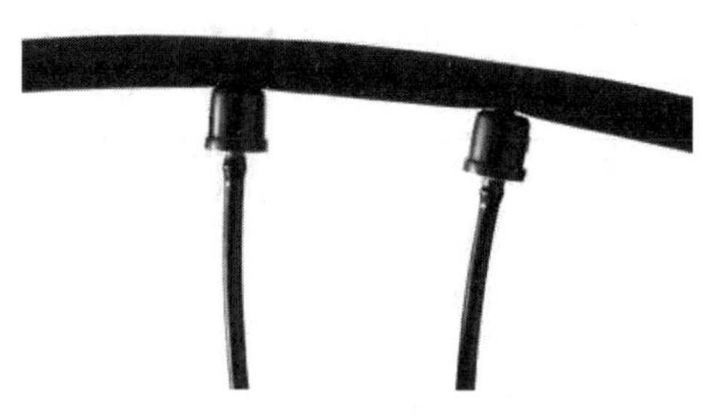

Figure 5-10 Basic structure of tubule outflow device

(2) Regulator

① Structure

The regulator is a miniaturized pressure compensator, functioning to keep the flow rate of tubule outflow device constant within a certain range of pressure. The basic structure of the regulator is similar to the pressure compensation emitter.

② The main technical indexes and parameters

The main technical indexes and parameters of the regulator are the same as the pressure compensation emitters.

(3) The relationship between flow rate and working pressure of the tubule outflow device

The relationship between the flow rate and the working pressure of the tubule outflow device is the same as that of other micro-irrigation emitters. However, because the tubule outflow device is composed of two different PE plastic tubules, its hydraulic characteristics are a synthesis of the small tube and the regulator.

① Relationship between the flow rate and the working pressure of the tubule

The use of a $\phi 4$ PE plastic tubule is the most suitable structure for a tubule outflow device. According to the calculation below, the relationship between the flow rate and pressure of a $\phi 4$ PE plastic tubule can be expressed as follows:

$$h_f = 585 \times 10^{-6} q^{1.733} L \tag{5-5}$$

where q is the flow rate of the tubule outflow device in L/h, h_f is the head loss of tubule outflow device in m, and L is the length of the $\phi 4$ tubule in m.

② Relationship between the flow rate and the working pressure of the tubule outflow device

At present, the data of different rates of the flow rate and the working pressure of the regulator put forward by domestic manufacturers are obtained under the condition of atmospheric outflow. However, when regulators and the tubules composed of tubule outflow devices are used in an application, the pressured irrigation water flows through the regulator from the lateral into the tubule, it's an extrusion flow. Therefore, the relationship between the flow rate and the working head of the regulator cannot be directly described by the manufacturer, which must be corrected.

When the length of the tubule is determined, the water head loss of the tubule can be calculated. The relationship between the flow rate and the head of the tubule outflow device is obtained from calculating the correlation between the flow rate and the head of the emitter:

$$q = k(h + h_f)^x \tag{5-6}$$

where q is the flow rate of the tubule outflow device in L/h, k is the flow rate coefficient of the regulator, h is the working head of the regulator in m, h_f is the head loss of the ϕ4 PE tubule in m, and x is the flow index of the regulator.

In a situation where the flow rate is less than 30 L/h and the length of the tubule is less than 0.5 m, the water head loss caused by the tubule is often overlooked due to how small it is.

3. Minisprinkler

(1) Types of minisprinklers

① Refraction type

The refraction minisprinkler is mainly composed of three parts: a nozzle, refraction cone and bracket. Pressured water jets through the nozzle, impacting the refraction cone, forming the water film, and spraying around. The basic structure of the refraction minisprinkler is shown in Figure 5-11. This kind of minisprinkler has a simple structure, which is advantageous as it has no moving parts, reliable operation, and a low price. Unfortunately, this also means the wetted radius is small.

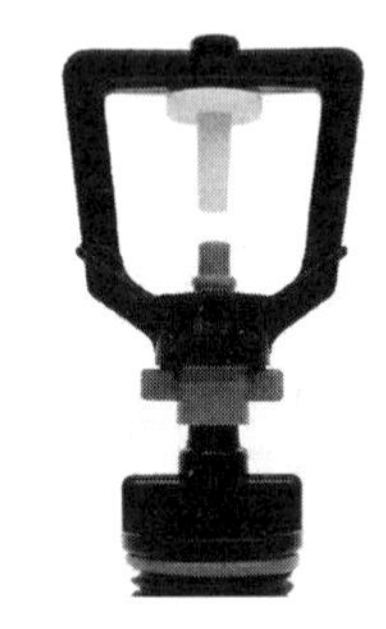

Figure 5-11 Refraction minisprinkler

② Rotator minisprinkler

The rotator minisprinkler is mainly composed of a rotating refraction arm, nozzle and bracket. The pressured water flows through the jet body, injected by the nozzle to impact the refraction arm. The flow forces the refraction arm to rotate and is subjected to the counterforce of the refraction arm to spray outwards at a certain elevation angle. Figure 5-12 shows two different structural forms of this type of minisprinkler. Its large wetted radius and low water application rate are beneficial, but the moving parts require a high manufacturing accuracy and the system is easily worn down, affecting the service life.

In addition to the non-pressured compensation function, the minisprinkler also has the function of preventing dripping, which is suitable for the suspension use of a greenhouse, meeting the needs of different occasions.

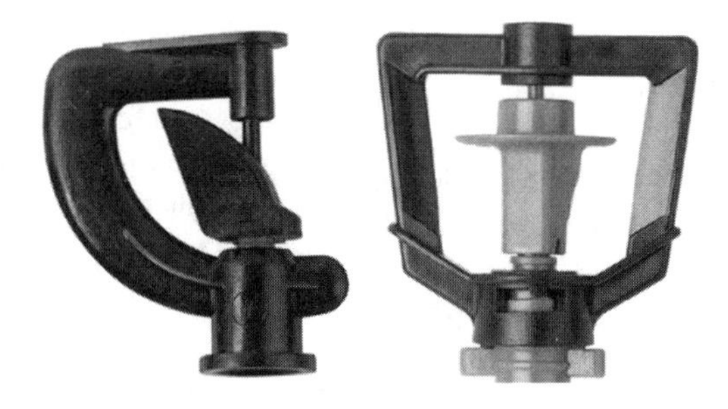

Figure 5-12 Rotator minisprinkler

(2) Main technical indexes and parameters

According to the GB/T 18687—2002, the main technical indicators of the minisprinkler are as shown in Table 5-4.

Table 5-4 Main technical specifications of the minisprinkler

<table>
<tr><th colspan="2">Project</th><th>Qualified index</th><th colspan="2">Test conditions</th></tr>
<tr><td colspan="2">Rated flow rate/(L/h)</td><td></td><td rowspan="6">Water temperature (23 ± 2)℃, filter screen has 160 to 180 mesh; The diameter of the mesh is less than the diameter of the nozzle diameter</td><td rowspan="4">Pressure range is 0.5 to 1.5 times the rated working pressure</td></tr>
<tr><td colspan="2">Working pressure range/kPa</td><td></td></tr>
<tr><td rowspan="3">Hydraulic performance</td><td>Flow index x</td><td></td></tr>
<tr><td>Flow rate coefficient k</td><td></td></tr>
<tr><td>Wctted radius R</td><td></td><td rowspan="2">Working pressure 25 kPa</td></tr>
<tr><td>Structure performance</td><td>Manufacturing deviation coefficient C_v</td><td>>7%</td></tr>
<tr><td colspan="2">Deviation rate between average flow rate and rated flow rate C</td><td>The non-pressure compensation type is less than 7%; Pressure compensated type is less than 10%</td><td></td><td></td></tr>
<tr><td colspan="2">Durability</td><td>Continuous operation of 1 500 h without fault, no visible defects, and the initial flow deviation ≤|±0.1|</td><td>Rated operating pressure</td><td></td></tr>
<tr><td colspan="2">Drawing resistance</td><td>In axial connection, the axial tensile force at both ends of the specimen is maximum in the 30 s without damage or detachment. There is no damage and no separation in the 30 s when the pulling force on the pipe is 30 N</td><td>At normal temperature, connect with PE plastic pipe according to the connection mode provided by the manufacturer</td><td></td></tr>
</table>

The technical parameters of the minisprinkler provided by the manufacturer should generally include the rated working pressure (kPa or m), rated flow rate (m^3/h), relationship between the wetted radius, flow rate and working pressure under the rated working pressure (or flow index x and flow rate coefficient k), and manufacturing deviation coefficient C_v. The performance parameters of several minisprinklers are shown in Table 5-5.

Table 5-5 Technical parameters of several minisprinklers

Name/ Model	Nozzle diameter/ mm	Working pressure/ kPa	Nominal flow rate or Rated flow rate/ (L/h)	Spray diameter/m	Relationship between flow rate and working pressure $q=kh^x$
RONDO refraction mini-sprinkler	0.8	300	47	2.0~2.4	$q=11.16h^{0.4132}$
	1.0	300	61	2.0~2.4	$q=12.786h^{0.4870}$
	1.2	300	91	2.5~2.8	$q=15.755h^{0.5157}$
RONDO XL rotator mini-sprinkler	1.6	200	135	7.0	$q=30.86h^{0.4918}$
	1.8	200	170	7.0	$q=41.2h^{0.477}$
	2.0	200	210	7.0	$q=50.169h^{0.4807}$
	2.2	200	260	7.0	$q=63.523h^{0.4712}$
	2.4	200	305	7.0	$q=73.372h^{0.4768}$
	2.6	200	367	7.0	$q=82.559h^{0.4977}$
W1507PT rotator mini-sprinkler		180	61	7.2	$q=0.5332h^{0.47}$
		280	75	8.4	

4. Spray hose

The spray hose is an emitter that many holes on a plastic hose at a certain distance sprays water onto a certain arrangement of land. Water is discharged through the water outlet to trickle or spray onto the soil. Depending on how the spray hose is set up and used, the designated flow rate from the emitter can be divided into two categories. A water outlet flow rate of less than 250 L/h in a single group is sprayed using a micro-irrigation hose, and a water outlet flow rate greater than 250 L/h is sprayed using a sprinkler irrigation hose.

The spray hose manufacturing process is simple, flexible, and convenient, as it is not costly, and suitable for vegetables, fruit trees, tea gardens and all kinds of field crops, especially scattered plots.

(1) Types and structures

① Drip type

A single set of water outlet flow rates less than 15 L/h with a drip belongs to the drip type. This kind of spray hose can be used in combination with plastic film mulch, making it similar to drip irrigation under mulch. The diameter of this kind of spray hose is generally from 20 to 32 mm, the aperture of it is from 0.5 to 1.2 mm, and it has 2 to 3 single hole groups.

② Microspray type

A single set of water outlet flow rates greater than 15 L/h and less than 250 L/h with a water spray is known as a microspray hose. This kind of spray hose is commonly used in the irrigation of open vegetable fields, fruit trees, and greenhouses.

The diameter of this kind of sprayhose is generally 40 mm, the aperture ranges from 0.5 to 0.9 mm, and it has 3 to 5 single hole groups. Depending on the condition of an area and the different necessities, a microspray can be classified as an ordinary type, edge-pressing type, and reinforced edge-pressing type. The edge-pressing type, also called a double wing, can make the water jet hose work without turning and moving out of its exact position. The reinforced edge-pressing type is based on the edge-pressing type as an electric machine, and a plurality of reinforced ribs are longitudinally added on the upper and lower walls of the hose to meet the requirements of higher water pressure.

③ Sprinkler irrigation type

A single set of water outlet flow rate greater than 250 L/h with a flow spray is classified as a sprinkler irrigation hose. This kind of spray hose is widely used for irrigating vegetables, tea gardens, and fruit trees in an open field. The working condition of the sprinkling irrigation hose is shown in Figure 5-13.

Figure 5-13 Application of the sprinkler irrigation hose

The structure of this kind of spray hose is similar to that of the microspray, but the diameter and thickness of the hose walls are larger.

(2) Technical indexes and parameters

The technical standards of a water spray hose can be implemented according to the NY/T 1361—2007 industry-standard issued by the Ministry of Agriculture in 2007. The standards require the main technical parameters of the spray hose should be included in consideration of the actual needs of the field or area to be irrigated. This includes but is not limited to the diameter, the flow rate of a single hole group (nominal or rated), hole distance, rated working pressure, wetted width, 100 mm spray uniformity coefficient, and bearing capacity. At present, the technical parameters provided by spray hoses in the market are far from meeting the requirements. Table 5-6 lists the technical parameters of several spray hoses developed by Beijing Wing Ring Energy Technology Co., Ltd.

Table 5-6 Main technical parameters of typical spray hose

Model	Specifications pipe diameter×thickness/mm	Working head /m	Relationship between working pressure and flow rate	Maximum service length/m (q_v=15%)	Maximum wetted width/m	Blasting pressure head/m
Single-hole drip irrigation hose	20×0.2	1~3	$q=4.91h^{0.56}$	≤80	0.6	12
Double-hole drip irrigation hose	20×0.2	1~3	$q=9.81h^{0.56}$	≤50	0.8~1.0	12
Double-hole drip irrigation hose	25×0.15	1~3	$q=11.19h^{0.56}$	≤70	0.8~1.0	9

Continued

Model	Specifications pipe diameter × thickness/ mm	Working head /m	Relationship between working pressure and flow rate	Maximum service length/m (q_v=15%)	Maximum wetted width/m	Blasting pressure head/ m
Three-hole drip irrigation hose	20×0.2	1~3	$q=14.73h^{0.56}$	≤40	0.8~3.0	12
Three-hole microspray hose	40×0.3	3~5	$q=27.24h^{0.56}$	≤100	3~5	15
Five-hole microspray hose	40×0.3	3~5	$q=27.67h^{0.56}$	≤100	3~5	15
Five-hole microspray hose	40×0.4	3~6	$q=24.93h^{0.56}$	≤100	3~6	20
Seven-hole microspray hose	40×0.4	3~5	$q=27.95h^{0.56}$	≤100	3~5	15

5.4 Pipelines and fittings

The pipeline is the main component of the micro-irrigation system. Various pipelines and fittings are assembled into a micro-pipe transmission and distribution network according to the design requirements, and the water is distributed to fields and crops in a manner that meets the specific crop's water requirements. Pipelines and fittings in large quantities of micro-irrigation projects have unique specifications. The amount of investment put into the quality is significant as the quality of the pipelines and fittings is not only directly related to the amount the project will cost, but also to ensuring a functional operation to properly irrigate the land, and how long the system can work.

5.4.1 Basic requirements for pipelines and fittings used in micro-irrigation system

1. Sustaining pressure

The micro-irrigation pipeline network is pressured; all levels of the network must be able to withstand a designated pressure to ensure safe transport and distribution of water. Therefore, the bearing capacity of all kinds of pipelines and fittings must be known when selecting pipelines for a system. The bearing capacity of pressure is directly related to the material, specification, type and connection mode of the pipeline and the fitting. Therefore, it is necessary to understand the material and properties of a pipeline and its fittings so as not to affect the quality of the project.

2. Strong corrosion resistance and age resistance

In a micro-irrigation system, the emitter orifice is very small. Therefore, the micro-irrigation pipeline network requires that the pipelines and fittings have strong corrosion resistance to avoid corrosion, precipitation, microbial reproduction, and other clogging emitters in the process of water delivery and distribution. Pipelines and fittings should also have a strong age resistance. For plastic pipelines and fittings, a certain proportion of carbon black must be added to improve anti-aging properties.

3. Specifications and tolerances must conform to technical standards

The deviation of the pipeline diameter and wall thickness should be within the allowable range of technical standards, and the inner wall of the pipeline should be smooth to reduce head loss. The pipeline wall, to be smooth, should have no dents, cracks, or bubbles, and the connectors should have no bubbles or burrs.

4. Low price

Micro-irrigation pipelines and fittings account for a large proportion in the investment of the entire system and should strive to meet both micro-irrigation project quality requirements as well as not being too costly.

5. Easy construction and installation

The connection between various fittings and pipelines should be simple, convenient and watertight.

5.4.2 Types of micro-irrigation pipelines

Undergoing micro-irrigation projects should utilize plastic pipes. Large scale micro-irrigation projects should specifically utilize pipelines. If the plastic pipes cannot meet the specific design requirements of the project, other materials can be used in the pipeline, as long as they can prevent rust clogging in the emitters.

There are two kinds of plastic pipes commonly used in a micro-irrigation system: PE pipes and PVC pipes. Polyethylene pipes are used in pipes under 63 mm, and PVC pipes are used in pipes greater than 63 mm. Plastic pipes have the advantages of corrosion resistance, good flexibility in adapting to small local settlements, smooth inner walls, little roughness in the water conveyance process, lightweight, convenient transportation, and installation, etc. It is an ideal micro-irrigation pipe. The main disadvantage of plastic pipes is that they are susceptible to aging by sunlight. When the plastic pipe is buried in the ground, the aging of the plastic pipe will be greatly delayed, and the service life can be over 20 years.

1. Polyethylene (PE) pipe

PE pipes are divided into two types: high-pressure low-density, and low-pressure high-density. The low-pressure high-density PE pipe is rigid, and the pipe wall is thin. The high-pressure low-density PE pipe is a semi-hose, which has a thicker tube wall and stronger adaptability to the terrain than low-pressure high-density PE pipes.

High-pressure PE pipes are made of LDPE resin, stabilizer, lubricant, and a certain proportion of carbon black. They have a better high impact and low-temperature resistance, lightweight, durability, and anti-aging properties than polyethylene pipes, but are not wear and high temperature resistant, nor have low tensile strength.

To prevent sunlight from entering the pipe through the pipe wall, algae and other microorganisms are bred in the pipe to absorb ultraviolet radiation, slow down the aging process, and enhance the anti-aging properties. PE pipes should be black and smooth both externally and internally having no bubbles, cracks, grooves, depressions, or impurities.

2. Polyvinyl chloride (PVC) pipe

The PVC pipe is mainly made of PVC resin, forming together with stabilizer, lubricant, and so on. It has good impact resistance, rigidity and bearing capacity, but poor resistance to high temperature. Softening deformation occurs when the temperature is above 50 degrees centigrade. PVC pipes are hard tubes with tough durability, but its adaptability to any given terrain is inferior to semi-flexible, high-pressure PE pipes. PVC pipes used in micro-irrigation are usually grey. To ensure the quality, the internal and external walls of the pipe should be smooth, not being corrugated or having bubbles, cracks, or indentation. The deflection of the pipe with a diameter of 40 mm to 200 mm should not exceed 1%. The thickness deviation of the same section of the pipe should not exceed 14%. PVC pipes are classified as light and heavy depending on the pressure used. Light water pipes are often used in micro-irrigation systems, that is, when the water pressure under normal temperature is no more than 600 kPa. The length of each pipe is generally from 4 to 6 m.

5.4.3 Types of fittings for micro-irrigation pipelines

Fittings are components that link pipes together, also known as pipeline fittings. Different types of pipelines have different fittings. In the majority of micro-irrigation projects using polyethylene pipes, this chapter only introduces PE pipe fittings. For PVC pipe fittings, please refer to the sprinkler irrigation section. At present, the domestic use of PE plastic pipe fittings in micro-irrigation has two categories. One is the Beijing Luyuan Company representing external fittings, which are below 20 mm with internal fittings. The second is the Shandong Laiwu Plastic Products Factory representing inscribed fittings. When users choose between these two kinds of fittings of different sizes, they must understand the correlation between the fittings and the pipelines.

1. Joint

The function of the joint is to connect the pipes. Depending on the size of the two connecting pipes, it is divided into the same diameter and different diameter joints. The PE joints are divided into three types based on their mode of connection: inverted hook-socket inserted joints, threaded joints, and threaded socket inserted joints.

2. Tee

Tee is the connector used for the differential of the pipeline, like joints. There are equal diameter tees and unequal diameter tees. Each structure hast wo connection types, which are either inverted hooks or thread connections, as shown in Figure 5-14.

Figure 5-14 Different types of tee

3. Elbow

Where terrain slope change is bigger in a given irrigated area, and where there are curve turnings, it is necessary to use elbow connections. The structure of elbow connections has three types: inverted hook interpolation, threaded connection, and thread locking connection.

4. Plug

The plug is used to seal the end of the pipe. It is displayed in Figure 5-15.

5. Bypass

The bypass is used to connect the lateral to the branch pipe. Its structure is shown in Figure 5-16.

Figure 5-15 Plug

Figure 5-16 Structure of the bypass

6. Inserted link

The inserted link is used to support the minisprinkler, so that the minisprinkler is placed at the specified height and has different forms and heights, as shown in Figure 5-17.

7. Sealing fasteners

The fastening internal fittings are shown in Figure 5-18.

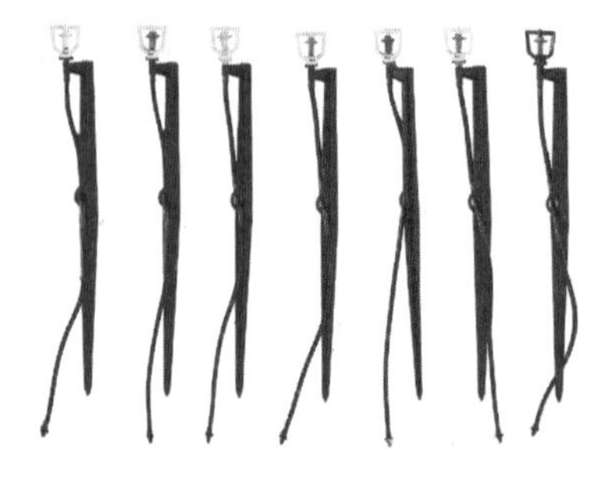

Figure 5-17 Inserted link

Figure 5-18 Seal ring

5.5 Control, measurement and protection devices

To control the micro-irrigation system and ensure its normal operation, the necessary control, measurement, and protection devices must be installed in the system. These devices include valves, flow rate, pressure regulators, flow meters or pumps, pressure gauges, safety valves, and air release valves.

5.5.1 Air release valves

The air release valve can automatically exhaust and suction air, and when the pressured water flows through, it can automatically shut down. In micro-irrigation systems, it is mainly installed in the highest levels of the pipeline network system. When the pipeline begins to deliver water, the air in the pipe begins to be pushed out of the water to the height of the pipeline. When the air cannot be discharged, it will reduce the water cross-section, and cause pressure shocks many times higher than the working pressure by several times. Exhaust valves should be installed at these high points so that the air in the pipe can be discharged in time. When the water is stopped, water will still be flowing through the high pipe because the water in the pipeline is gradually discharged to the lower level. The air release valve can make the air in time so that the air can enter the pipeline.

5.5.2 Flow rate and pressure regulator

Flow rate and pressure regulators are devices for automatically regulating the pressure and flow rate in a pipeline.

1. Flow rate regulator

The flow rate regulator adjusts the flow by automatically changing the size of the cross-section. Under normal working pressure, the rubber ring in the flow rate regulator is in a normal working state, and the flow rate is at the designated value. When the water pressure increases, the water pressure keeps the flow rate constant, and thus ensuring the stability of the flow rate in the pipeline at all levels in micro-irrigation systems.

2. Pressure regulator

A pressure regulator is used to regulate the water pressure in a micro-irrigation pipeline while stabilizing it. The safety valve is also a special pressure regulator. The working principle of the pressure regulator used in a branch pipe or lateral inlet of a micro-irrigation system is to adjust the pressure in the pipe by changing the cross-section of the water by the force and deformation of the spring, so that the pressure at the outlet of the pressure regulator remains stable (in fact, it is also a flow rate regulator).

3. Measuring device

(1) Pressure gauge

The pressure gauge is an essential measuring instrument in a micro-irrigation system. It can reflect whether the system is designed according to its normal operation, especially before and after the application of a filter pressure gauge. It is a reflection of filter clogging degree and indicator when cleaning filters.

The pressure measuring device commonly used in micro-irrigation systems is the bourdon tube pressure gauge, which is a spring tube with a circular cross-section fixed in the pressure gauge. One end of the pipe is fixed in the socket and connected with the external joint, and the other end is closed and connected with the connecting rod and the sector gear, which can move freely.

(2) Water meter

In a micro-irrigation system, the total amount of water flow through a pipe or the amount of water irrigated in a period can be measured by a water meter. A water meter is generally installed on the main pipe behind the filter in a headwork pivot, and a water meter can be installed on the corresponding branch pipe.

(3) Electromagnetic flowmeter

The electromagnetic flowmeter (Figure 5-19) can accurately measure the flow rate of all conductive liquids without mechanical obstruction and has various insulation armor, lining, and signal converters, which can be used to combine the flowmeter suitable for various occasions.

(4) Ultrasonic flowmeter

The ultrasonic flowmeter (Figure 5-20) can be used to measure the flow rate of non-conductive liquid accurately and without mechanical hindrance. The design makes it suitable for various installation conditions. The flowmeter is an integral structure; the sensor and the tube wall are integrated. The digital signal sensor has lasting stability and can be installed on an existing pipeline without water supply.

Figure 5-19 Electromagnetic flowmeter

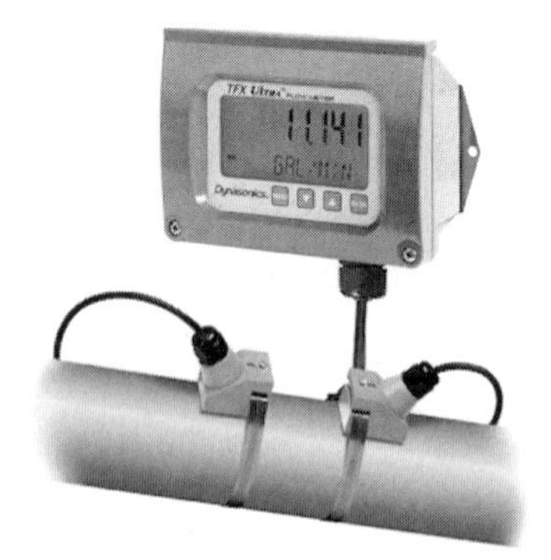

Figure 5-20 Ultrasonic flowmeter

5.6 Filter equipment

Filtration refers to the process of removing suspended solids in irrigation water. The device used is generally called a filter, which is one of the main components of the micro-irrigation system. Because the diameter of the irrigation emitter is small (generally from 0.8 to 2.0 mm), the filter is an essential part of the micro-irrigation system. For a sprinkler irrigation system, because of the large sprinkler tube (usually more than 1.2 mm), the installation of filters depends on the sprinkler flow rate and water quality. In most cases, there is generally no need to install filters. When the water source is an open channel or a pond, the sprinkler irrigation system can also install the corresponding filter to prevent algae, silt clogging, and damaging of the sprinkler. The common filters are screen filters, sand filters, laminated filters, and hydro-cyclone water and sand separators. The combination filter is usually used in engineering, and only in a few instances can single filters be utilized. Also, when the quality of irrigation water is poor, it is necessary to add a primary filter device upstream of the filter, also known as a pre-filter device. In this section, the structure, principle and application conditions of the above filters are introduced, and the selection methods of the combined filter device and the pre-filter device are introduced combined with the actual engineering application.

5.6.1 Screen filter

1. Basic structure

A screen filter is a kind of filter device with a simple structure and universal application. Its basic components include a screen cloth core (filter element) and filter shell two parts. The structure schematic diagram is shown in Figure 5-21.

Figure 5-21 Structure schematic diagram of the screen filter

The screen filter element is made up of a skeleton and screen cloth. The skeleton is a cage-like structure which supports the filter. It can bear a certain lateral pressure without deformation, and both ends have a sealed rubber ring, which can be closely matched with the filter shell. The screen cloth generally requires more than 160 mesh (0.1 mm in diameter). The screen should be smooth and uniform bonding or welding in the surface of the skeleton. The filter must be able to resist corrosion and be easy to change out and clean. The filter shell is a device with the filter element in it, consisting of an upper and lower parts. The upper part is usually called the gland, and the lower part is the main shell. The filter shell is equipped with a water inlet, a water outlet, and a sewage outlet. The water inlet is in conjunction with the inner cavity of the filter element, while the outlet is arranged on the side of the main shell. The filter shell material can be made of either steel or plastic; the flow of the water is what determines the best material for each filter. A small flow rate ($\leqslant 9\ m^3/h$) filter utilizes the plastic shell, and a large flow rate filter ($> 9\ m^3/h$) adopts the steel shell. In recent years, with the development of the plastic industry, the large flow rate plastic filter has gradually replaced the use of the steel filter. Whether it is steel or plastic filter, though, the filter element must be made of corrosion-resistant materials.

2. Working principle and characteristics

The screen filter works by using the mechanical sieving function of the screen to intercept the solid suspended particles whose particle size exceeds the pore size of the mesh to filter irrigation water. The raw water flows from the water inlet to the inner cavity of the filter element, and the larger particles inside the water are caught from passing through the screen. Particles that are smaller than the net aperture pass through the screen and enter the irrigation pipeline network with the filtered water from the water outlet. The efficiency of the screen filter mainly depends on the screen size used; the larger the mesh of the screen cloth, the higher the filter precision.

Screen filters are one-dimensional plane filtrations, which effectively filtrates granular suspended solids, but has poor filtration when it comes to filamentous, linear particles and latex particles. The filter elements of the screen filter are usually washed by hand. When the pressure on both sides of the filter increases to the specified index, the filter needs to be stopped and opened to remove the blocked areas of the filter. Then, the spare filter is replaced after one manually brushes and cleaned all of the blocked entries of the mesh.

3. Technical parameter

(1) The screen cloth

The screen cloth is used to precisely filter out particles from irrigation water, but the twine diameter in each type of screen cloth can inhibit the accuracy of the filtering process. Therefore, the mesh size and twine diameter are considered when using a filter so that it is the most accurate mesh for the specific filtration process. The conversion formula for mesh number and mesh diameter is as follows:

$$M=1/(D+\alpha) \tag{5-7}$$

where M is the mesh number, D is the mesh diameter (mesh length) in inches, and α is wire diameter in inches.

The partial screen cloth structural parameters and controls are shown in Table 5-7.

Table 5-7 Structure parameters of the industrial woven wire cloth with square holes

Mesh aperture/mm	Wire diameter/mm	Percentage of sieving area/%	Net weight per unit area/ (kg/m²)			Equivalent imperial number (Mesh/25.4 mm)
			Brass	Tin bronze	Stainless steel	
1.00	0.500	44.4	2.35	2.38	2.14	16.93
0.900	0.500	41.3	2.51	2.55	2.30	18.14
0.800	0.450	41.0	2.28	2.31	2.08	20.32
0.790	0.450	37.5	2.46	2.49	2.25	21.90
0.600	0.400	36.0	2.25	2.29	2.06	25.40
0.500	0.315	37.6	1.71	1.74	1.57	31.17
0.400	0.250	37.9	1.35	1.37	1.24	39.08
0.300	0.200	36.0	1.13	1.14	1.03	50.80

Continued

Mesh aperture/ mm	Wire diameter/ mm	Percentage of sieving area/ %	Net weight per unit area/ (kg/m^2)			Equivalent imperial number (Mesh/25.4 mm)
			Brass	Tin bronze	Stainless steel	
0.200	0.120	34.6	0.81	0.82	0.74	74.71
0.150	0.900	36.0	0.56	0.57	0.51	91.60
0.125	0.090	33.8	0.53	0.54	0.45	118.14
0.900	0.080	30.9	0.50	0.51	0.46	141.11
0.080	0.063	31.3	0.39	0.40	0.36	177.62
0.071	0.056	31.3	0.35	0.35	0.32	200.00
0.063	0.050	31.1		0.32	0.28	224.78
0.050	0.040	30.9		0.25	0.23	282.22
0.040	0.036	27.7		0.24	0.22	334.21

Note: From GB/T 5330—2003.

(2) Filter parameter

① Cleaning pressure drop. The head loss between the inlet and outlet of the filter without a filtration load is called a cleaning pressure drop. Generally, the cleaning pressure drop is less than 30 kPa, and the flushing pressure difference is less than 50 kPa. If the pressure difference is too large, the filter element will deform easily.

② Flow rate. The maximum filter flow rate is 0.14 $m^3/(s \cdot m^{-2})$, and the moderate flow rate ranges from 0.028 to 0.068 $m^3/(s \cdot m^{-2})$.

③ The net pore area should be 2.5 times larger than the discharge area of the outlet pipe.

(3) Flushing water consumption

Hydraulic backwashing common screen filters are ineffective. It is usually better to replace the filter, or manually wash and rinse it. The water consumption of the filter fluctuates depending on the circumstance. The productivity of the self-cleaning screen filter is related to the quality of the product. The cleaning time of a product with superior quality is about 9 s, and the water consumption is no more than 90 L.

4. Application condition and maintenance of the screen filter

The screen filter is suitable for micro-irrigation systems with good water quality, such as well water, tap water, and other clean water sources. At present, most of the drip irrigation systems used in greenhouses in China utilize small flow rate screen filters. The screen filters are usually installed downstream from sand filters or hydrocyclone separators in large and medium-sized irrigation systems. Screen filters upkeep the quality of the system by checking the working conditions of the filter regularly, cleaning or changing the filter regularly, and removing the filter after the irrigation period ends. Self-cleaning screen filters also need to check the condition and performance of the washing mechine.

5.6.2 Laminated filter

The laminated filter (Figure 5-22) has nearly 30 years of history in the field of irrigation application. At present, its performance has greatly improved, especially in the development and application of automatic flushing and multi-element composite laminated filters, which enhanced the promotion speed of the laminated filter in micro-irrigation engineering. At present, the laminated filter has been widely used in the world, and it tends replacing sand filters.

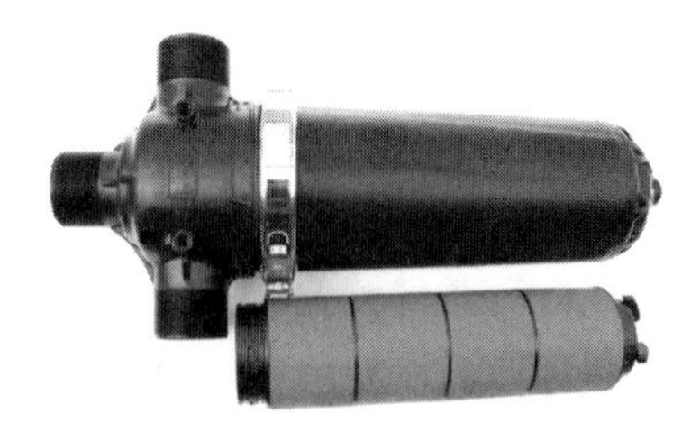

Figure 5-22 Laminated filter

1. Basic structure

The laminated filter is mainly composed of a shell, filter element and washing mechanism. The annular plastic sheet with a group of the microchannel is superimposed into a cylindrical shape and fixes the filter into a special support frame. The flushing mechanism is arranged inside the filter's porous flow holes, which are structured by numerous exhaust pipes.

2. Working principle

Plastic sheets that are stacked together form tiny holes and layers using numerous cross striations above them. When filtering, the raw water enters the small orifice of the filter from the outside, and large particles in the water are intercepted by the outside of the micro-hole or in the crevice of the lamination; this is how the irrigation water transitions from being raw to being filtered. After filtration, it flows into the inner cavity of the filter element, and then enters the irrigation pipeline network. When backwashing, the backwashing valve turns on and switches the direction of the water flow. Clean water enters the central disc, loosening the laminated extrusion ring and dispersing the plastic laminations. Impure particles on the disc are washed off with the force of the reverse flow jet, and the outlet is discharged with the flushing water flow.

3. Performance parameters of the laminated filter

(1) Filter precision

The effectiveness and precision of the laminated filter mainly depend on the groove size, shape, and stacking conditions on the plastic plate, usually to mesh or aperture. The common filter precision is 20 μm, 55 μm, 90 μm, 130 μm, 200 μm, 400 μm and so on.

(2) Filter flow rate and backwash flow rate

The filter flow rate can be defined by the nominal flow rate of a single (or unit) filter element. Generally, the flow rate of a single filter element ranges from 5 to 15 m^3/h, the backwash time is about 9 s, and the backwash water consumption ranges from 20 to 30 L. The laminated filter system is composed of many filter units (one unit with a filter element). A large filtration system can contain hundreds of filter units, and the filter flow rate can reach thousands of cubic meters per hour.

(3) Others

Other parameters include the minimum recoil pressure, system pressure loss, and water consumption of the backwashing unit. Here are the specific parameters of a group of brand-name products for designers, as follows:

Maximum operating pressure is 1 MPa; Minimum backwash pressure is 280 kPa; System pressure loss ranges from 8 to 80 kPa; Unit design flow rate is 5 m^3/h with filter precision of 20 μm, 9 m^3/h with filter precision of 55 μm, and 15 m^3/h with filter precision greater than 90 μm; Unit backwash flow rate ranges from 8 to 9 m^3/h; Unit backwash water consumption ranges from 17 to 33 L.

4. Conditions for use of laminated filters

Compared with sand filters, the laminated filters are the most advantageous. First, the structure is simple, light and low cost. The filter container is smaller, and the shell and filter elements use less raw materials, which saves costs and facilitates the transportation and installation of the filter. Second, the shell and filter elements are made of materials like stainless steel, which do not easily rust. The manufacturing precision is high, also helping the inhibition of rust as well as clogging. Third, less water is used in backwashing. However, the laminated filter has some disadvantages, such as a high washing frequency, high manufacturing precision requirements, and incomplete backwashing. The products of reliable, name-brand manufacturers should be selected.

The filtration effect of the laminated filter is similar to that of the sand filter, and it is effective in filtering inorganic and organic suspended particles. It is generally used as the main filter in an irrigation area with poor water quality.

With a variety of new materials and control technology, the application of the laminated filter has been extended to many new work fields, such as civil and municipal wastewater treatment centers, industrial wastewater processing centers, textile and steel factories, food processing industries, industrial cooling water, water treatment, and desalination centers, and many other areas.

Figure 5-23 Constitutional diagram of laminated filters

The form of a laminated filter and its combined structures are shown in Figure 5-23.

5.6.3 Sand filter

1. Basic structure

The sand filter is mainly composed of two parts: a shell and a filter bed. The shell is made of the steel pressure vessels. The most common filter materials are sand and gravel. The sand and gravel are accumulated in the cavity of the container to form a filter bed, forming a porous medium filter. Also, the sand filter also includes a backwashing mechanism, pressure and flow rate monitoring equipment, and water inlet and outlet pipes. The basic filter unit of the sand filter is a container; the structure of the container can be seen in Figure 5-24. The figure displays the inlet and outlet, support and sand hole, etc. on the outside of the container. The container contains a water distribution plate, filter material, filter head, water collecting tank (pipe) and filter bed support plate. The filter container has auxiliary facilities such as sand filling holes and inspection holes.

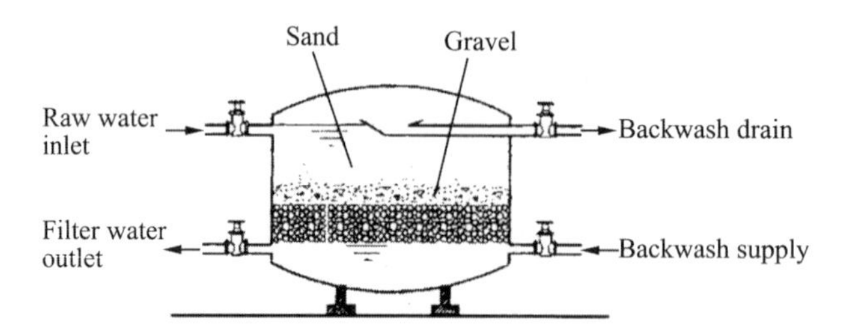

Figure 5-24 Schematic diagram of the sand filter container

2. Type

(1) Classification according to the structure of a filter bed

Depending on the structure of the sand bed, a sand filter can be divided into two types: graded sand filter and homogeneous sand filter. The homogeneous sand filter is commonly used in micro-irrigation systems because the backwash in micro-irrigation is frequent. When backwashing by the homogeneous structure, the phenomenon of hydraulic classification and surface filtration will not appear.

(2) Classification according to the installation method of the filter container

Based on the installation of the filter container, the sand filter can be either vertical or horizontal. The vertical filter has less filtration flow, usually below 90 m^2/h. The horizontal filter container is more economical when the large filter flow rate is needed.

(3) Classification according to work control mode

The sand filter is either in a manual backwashing working control mode or an automatic backwashing mode. Manual backwashing is the cleaning of filter materials by the manual operation when the filter works. An automatic backwashing filter is a special automatic control mechanism for the filter backwash operation.

3. Working principle

(1) Filtration principle

When the water flows through a filter bed with a certain thickness of quartz sand or granite grit, the suspended particles in the water are intercepted so that the irrigation water is clean and pure. This avoids blocking the irrigation pipeline and emitter. The accuracy of a sand filter depends on the particle size, the thickness of the filter layer, and the filtration rate of the sand filter material.

(2) Backwashing principle

Backwashing is the main way in which sand filters are cleaned. Backwashing encompasses the reverse flow from the lower part of the filter bed rinsing the filter. The filter material expands and forms a flow state, which will be fine suspended sediment in filter bed; impurities and sand particles and the like in the water is discharged to the outside of the filter. When the multi-tank combination filter backwashes, the majority of the container undertakes the filtering work. Part of the filtered water continues out to irrigate the field or another area to be irrigated. The other part of the water that does not irrigate the area forms a reverse flow for a few filter container washing cycles. Each filter container is washed in turn and flushed once for a flushing cycle and then returned to the normal filtration state.

4. Selection of quartz sand filter material

The sand filter material for micro-irrigation can be made of quartz sand or granite sand. As the granite sand often contains iron, manganese, and other metal elements, it may have adverse effects on the irrigation water. Therefore, quartz sand is used as the filter material in most cases. There are two sources of crushed quartz sand: a river or the original rock. River sand is naturally weathered along with the flow of migration. From there they become rounded particles with smaller porosity. Also, the river sand gets mixed with a variety of textures of sediment and soil and other substances; it must be washed to be used as filter sand. Artificial quartz sand is a kind of quartz ore that has been artificially crushed and therefore advantageously has a pure texture, many edges and corners, and a large porosity. The sand has various gradings and can be screened to ensure they pass the specifications and requirements. As it is a widely available, low price substance, it is convenient and used as the fundamental material of sand filtration systems.

In the selection of a sand filter, the relationship between the size of the emitter runner and the blockage should be considered. According to the data of the National Irrigation Engineering Manual of the United States, the commonly used sand filter models are No.11, No. 16, and No.20. The size of the suspended solid particles of No.11 granite sand (average particle size of 1 000 m) is larger than 80 m. The size of suspended solid particles of No.16 quartz sand (average particle size of 825 m) is larger than 40 m.

A quartz sand filter is mainly classified by two indexes: the average effective particle diameter and uniformity coefficient. 9% of all of the given sand particles must be smaller than the average effective particle diameter. For example, if the effective diameter of a kind of filter material is 0.8 mm, the diameter of 9% of these sand particles must be less than 0.8 mm. The uniformity coefficient is used to describe the diameter change rate of the sand filter material particle. It is the ratio of the diameter of 60% of these sand particles and 9% of these sand particles (d_{60}/d_9). A ratio equal to 1 means the filter material is the size of the particles. The uniformity coefficient of the filter material for micro-irrigation systems should be about 1.5.

The quality of filter material also depends on the chemical properties, shape, particle sizes, and other physical indicators of quartz sand. In the selection of filter materials, the filtration efficiency of the equipment, the cost of the material, and other factors listed below should be considered to make the most economical choice. The filter material needs to meet the following requirements:

① Sufficient mechanical strength to prevent wear and breakage of filter materials during filtration and backwashing.

② Sufficient chemical stability. Due to the use of fertilizer and pesticides during irrigation, and the chlorine treatment and acid treatment the system needs to prevent clogging, the filter material should not react with weak acid and weak base solution.

This could lead to a deterioration of the water quality, irrigation blockage, or the production of harmful material to plants and animals.

③ The particle size should be as uniform as possible, and a certain particle size distribution and appropriate porosity are necessary to ensure the uniformity coefficient is about 1.5.

④ Local materials are useful in reducing the cost of transportation.

The performance, physical, and chemical indexes of the quartz sand filter produced by a factory are listed in Table 5-8 and Table 5-9 respectively. These tables are useful as a reference when selecting quartz sand filter material.

Table 5-8 Performance index of quartz sand used in sand filters

Specification	Proportion/ (g/cm^3)	Porosity m/%	Spherical degree coefficient	Effective particle size d_9/mm	Uniformity coefficient
No.20	2.65	0.42	0.80	0.59	1.42

Table 5-9 Physical and chemical indexes of quartz sand material used in sand filters

Analysis project	Test data	Analysis project	Test data
SiO_2/%	≥99	Moh's hardness	7.5
Crushing rate /%	<0.35	Density /(g/cm^3)	2.66
Wear rate /%	<0.3	Bulk density /(g/cm^3)	1.75
Porosity /%	45	Boiling point /℃	2 550
Soluble hydrochloric acid /%	0.2	Melting point /℃	1 480

5. Basic parameters of the sand filter

(1) The thickness and particle size of filter material

The thickness of the filter material mainly constitutes the effectiveness and resistance of the filtration. The thicker the sand filter layer is, the more effective the filtration is, while the greater the water head loss will be. Experimental results show that a filter layer with the highest capability has a thickness of 30 cm or above, and the ability to filter when below 30 cm is very insignificant. Considering the influence of scour pits on the filter bed surface, it is necessary to increase the thickness to the safest level; the average filter thickness is about 50 cm.

The solid particle size that is to be filtered out of water is what determines the sand filter material. The particle size of solid suspension allowed to pass through the emitter channel of the micro-irrigation system is designed to be between 80 ~ 120 μm. For example, the suspended solid particle size that can be intercepted by No.16 quartz sand is 60 μm, and the value of No.20 quartz sand is 40 μm. They are therefore the most suitable for micro-irrigation system applications.

(2) Filtration parameters

① The flow rate and velocity of the filter. The filter flow rate refers to the water filtered in a period in m^3/h. Sometimes it can be represented by the unit time and unit area in $m^3/(H \cdot m^2)$. The velocity refers to the mean speed of the water flow through the filter layer in m/h. The flow rate is directly related to the velocity, and it increases with the increase of velocity. The flow rate range should be between 50～70 m/h in the practical application, meaning the flow rate will fluctuate from 50 to 70 $m^3/(h \cdot m^2)$.

② Pressure difference. Filter pressure indexes include the cleaning pressure drop and backwashing pressure drop. The cleaning pressure drop refers to the filter head loss value between an import and export under the rated flow in the unfiltered load conditions. It is an index of filter energy consumption, not usually greater than 30 kPa. The backwashing pressure difference refers to the head loss value when the filter layer needs cleaning. It is the control index of backwashing operation in the application of the filter.

(3) Backwashing parameters

① Backwashing speed. Backwashing speed refers to the strength of the process. The strength is determined by the flow rate or velocity of the reverse flow needed for the backwashing of the filter bed. It should be controlled within a moderate range; if the range is too high the filter bed will rush out, and if too small the strength will be not enough to flush. Generally speaking, the backwashing speed is related to the type of quartz sand used. The value of the No. 30 and No. 20 sand filter materials range from 24 to 36 $m^3/(h \cdot m^2)$, and the value of the No. 16 and No. 11 sand filters range from 48 to 60 $m^3/(h \cdot m^2)$.

② Backwashing time. Backwashing time is the amount of time required to clean the filter tank. The test shows that the backwashing time is not proportional to the backwashing speed. Backwashing time is generally controlled to be about 6 min.

③ Backwashing water consumption. Backwashing water consumption refers to the amount of clean water used in the process. It is expressed by the quotient of the total amount of water used by backwashing and the total water filtered. The percentage representing the backwashing water consumption rate of a sand filter ranges from 4% to 6%.

6. Use and maintenance of the sand filter

(1) Filter maintenance

The corrosion condition of the filter should be checked regularly. This includes the examination of the antirust coating internal and external to the filter, as well as the valve and pipe fittings. The sensitivity and accuracy of the control and monitoring equipment should be regularly checked, also while simultaneously recording and analyzing the reading value of each meter. Generally, maintenance and examination are needed before and after each irrigation cycle.

(2) Filter material maintenance

Filter materials need to be cleaned 2 to 3 times a year, depending on the water source used. Generally, the filter material should be replaced once every two irrigation seasons. After supplying or replacing the material, periodical cleaning is necessary to remove impurities carried by the filter material. When the filter bed surface is covered by algae or other organic substances, the "surface filtration" phenomenon happens, which greatly reduces the flow rate of the water and the filtration efficiency. At the end of the irrigation season, the water in the filter should be drained away, otherwise, it may cause clogging in the filtration material.

(3) Filtration water quality management

Sometimes, due to rainfall and other factors, the irrigation water's quality makes it unsuitable for use. In these situations, the filtration work of the filter should be stopped, so as not to cause damage to the filtration system and the entire irrigation system as a whole.

5.6.4 Self-cleaning filter

Self-cleaning filters can automatically clean the filter medium of the filter element. Self-cleaning filters come in three types: the screen filter, laminated filter, and sand filter. Self-cleaning screen filters rely on the automatic cleaning mechanism inside the filter to clean up the sundries that clog the screen surface. Self-cleaning laminated filters and sand filters use the automatic backwash cleaning mechanism to remove the blockage in the filter element or medium. In general, when it's necessary to clean the filter more than 3 times a day, it is best to use the self-cleaning filter. The three types of self-cleaning filters are examined below.

1. Self-cleaning screen filter

Self-cleaning screen filters and their structures are displayed in Figure 5-25. Its filtration principle is the same as an ordinary manual screen filter, while its cleaning mechanism is different. It works when the screen is blocked; the pressure difference between the water inlet and outlet increases. When the pressure difference reaches the set cleaning value, the automatic control device opens the blowdown valve, and simultaneously starts the cleaning mechanism inside the screen, so that it starts to rotate and walk. During the suction pipe rotary walking, the dirt plug on the screen cloth is sucked into the sewage pipe and the blockage is transported through it to the sewage outlet and finally is discharged to the outside of the filter by the drain valve.

The self-cleaning screen filter is characterized by its advantageous use of an adsorption cleaning method that has less cleaning water consumption, a shorter cleaning time, and a less harmful impact on the water supply during the cleaning process. However, it has limited filtration capabilities on organic matter.

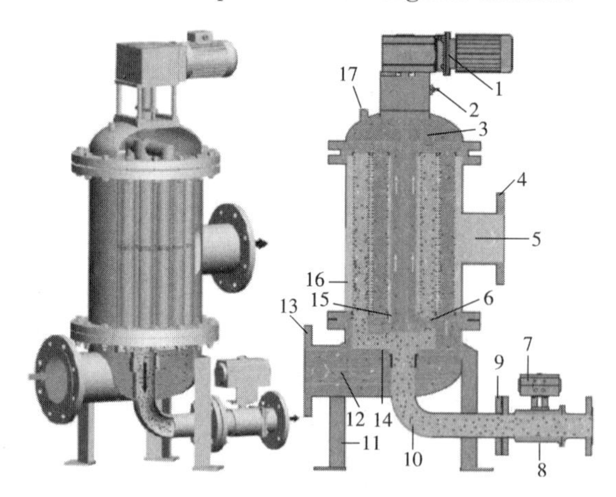

1—Gear motor; 2—Positioner; 3—Cover plate; 4—Outlet; 5—Filtrate; 6—Cleaning nozzle; 7—Electromagnetic switch; 8—Cleaning valve; 9—Discharging nozzle; 10—Discharging pipe; 11—Support leg; 12—Unfiltered liquid; 13—Inlet; 14—Cleaning arm; 15—Distribution pipe; 16—Filter element; 17—Vent

Figure 5-25 Self-cleaning screen filter

2. Self-cleaning laminated filter

Figure 5-26 Shape and structure diagram of the self-cleaning laminated filter

The filtration principle of the self-cleaning laminated filter is the same as that of the manual laminated filter, but the cleaning function is different. Self-cleaning laminated filters adopt the backwash mode. When backwashing, inflow water should be cut off from the filter, the drain valve opened, and the water flowing from the outlet into the internal of the filter. The laminations are dispersed, the reverse flow water filters out the blockage between the outside of the filter element and the laminations, and the water is then released through the discharge port. Its shape and structure are shown in Figure 5-26.

There are three common filter washing control indexes: ① the pressure differential controlled washing, washing when the import and export pressure differential exceeds a predetermined value. ② the timing washing, washing at a fixed interval. ③ the volume table controlled washing. When a predetermined amount of water passes through the filter, backwashing is initiated on the filter. The irrigation control index is determined mainly based on the project operation process, and sometimes on the testing site if necessary.

3. Automatic backwashing sand filter

The structure of the automatic backwashing sand filter is the same as that of the manual sand filter, and the only difference is that the manual operation is replaced by an automatic capability to initiate backwashing. The common washing control indexes are similar to those of stacked filters, including pressure differential controlled washing, timing washing, and volume table controlled washing. The process of filtration and backwashing of sand filters with the double container is shown in Figure 5-27. In the filtration state, two three-directional backwashing valves are opened at the same time, and the raw water is sent into the filter container. The water flows through the filter bed and into the irrigation pipeline network. Backwashing occurs in a three-directional function. The water inlet is cut off and the sewage outfall is opened, forming the reverse flow in the filter container, cleaning the filter bed, and finally releasing the dirt outflow through the sewage outfall. The automatic backwashing operation process is achieved by the three-directional backwashing valve controlled by an automatic controller.

Figure 5-27 Schematic diagram of filtration and backwashing process

5.6.5 Hydrocyclone water and sand separator

1. Basic structure and working principle

The hydrocyclone water and sand separator is a piece of common primary filtration equipment used in irrigation (shown in Figure 5-28). It operates when irrigation water containing sand is injected from the tangential inlet of the upper section at a certain flow velocity; as a result, a powerful rotation motion is formed in the hydrocyclone water and sand separator. Because the centrifugal force and fluid drag on the sand and water are different, more water is discharged from the overflow port through the internal circulating flow, while the sand and residual water move downwards along the separator wall, assembling at the bottom of the dirt collecting box, and are filtered regularly.

Figure 5-28 Schematic diagram of hydrocyclone water and sand separator

The percentage of sand grains that can be separated and removed by the hydrocyclone water and sand separator can reach 98% in a 200 mesh screen filter, but only when the specific gravity of the separated particles is higher than that of water. When the gravity is higher, the particles can be effectively removed, but if the gravity is lower, organic impurities cannot be removed. Also, the swirl generates a high head loss and consumes energy rapidly, so it is generally used as a first-class processing equipment of the filtration system, installed near wells and pumping stations. It is the most suitable equipment for water sources such as wells or rivers that contain a large amount of sediment.

2. Technical performance parameters and management maintenance

The technical parameters of the hydrocyclone water and sand separator include the minimum cleaning pressure drop, the minimum flow rate, the range of flow rate, and the rated working pressure. The minimum cleaning pressure drop refers to the minimum allowable pressure difference between an inlet and outlet. The minimum flow rate refers to the required minimum flow in the process of separating water and sand; if the flow rate or pressure difference is too small, the effective centrifugal flow won't form, therefore leading to sand particles going undetected for filtering. The working flow rate range and the rated working pressure are the design parameters the manufacturer must provide. The parameters of a hydrocyclone water and sand separator produced by a specific company are shown in Table 5-10.

Table 5-10 Technology parameters of a hydrocyclone water and sand separator produced by a company

Model	Minimum flow rate/ (m^3/h)	Maximum flow rate/ (m^3/h)	Size of inlet and outlet/ mm	Dimension/ mm	Weight/ kg
50	5	20	50	550×750	30
80	9	40	80	450×1 250	55
90	30	80	90	600×1 460	80

The maintenance of the hydrocyclone water and sand separator is mainly intended to prevent internal corrosion and external mechanical damage. Sediment particles and impurities should be regularly removed during usage.

5.6.6 Combination filter

In micro-irrigation systems, most of the filter systems utilize a combination of the main filter and a secondary filter. During usage, the water quality of the irrigation source is determined first, and the type and combination mode of the filter are selected to match according to emitter.

1. Selection of filter types

Each type of filter is most effective at removing a different type of contaminant in irrigation water. The corresponding filters can be selected about their effective filtration degree for various kinds of contaminants. Depending on the local economic conditions of a given area, filters should be selected that will be the most practical to each situation. In general, the sand filter is more expensive, while the segmented filter or screen filter is cheaper. The self-cleaning filter can be used if the economic conditions permit (shown in Table 5-11).

Table 5-11 Selection of filter types

Dirt type	Pollution degree	Quantitative standard/ (mg/L)	Swirl sand separator	Sand filter	Segmented filter	Screen filter
Soil particles	Low	≤50	A	B		C
	High	>50	A	B		C
Suspended solids	Low	≤50		A	B	B
	High	>50		A	B	C
Algae	Low			B	A	C
	High			A	B	C
Iron oxide and manganese	Low	≤0.5		B	A	A
	High	>0.5		A	B	B

Note: the controlled filter refers to the two-stage filter in a field. A is the first option, B is the second option, and C is the third option.

2. Selection of combination filters

The filter combination should be selected according to the quality condition of the irrigation water. The data below references different filter types and combinations recommended depending on the concentration and the size of impurity particles in water (Table 5-12).

Table 5-12 Reference table for selection of combination filtration systems

Water quality			Types and combination of filters
Inorganic substance	Content	<9 mg/L	Screen filter (laminated filter) or sand filter add the screen filter (laminated filter)
	Particle size	<80 μm	
	Content	9～90 mg/L	Hydrocyclone water and sand separator add the screen filter (laminated filter) or hydrocyclone water and sand separator add the sand filter and screen filter (laminated filter)
	Particle size	80～500 μm	
	Content	>90 mg/L	Sedimentation tank add the screen filter (laminated filter) or sedimentation tank add sand filter and screen filter (laminated filter)
	Particle size	>500 μm	
Organic compound	<9 mg/L		Sand filter add the screen filter (laminated filter)
	>9 mg/L		Trash rack add sand filter and screen filter (laminated filter)

3. Installation sequence of the combination filter

In general, the installation order of filters going from upstream to downstream is as follows: hydrocyclone water and sand separator, screen filter (or sand filter), laminated filter, and screen filter. When the self-cleaning screen filter is used alone, the primary filtration facilities are needed upstream. If the water quality is poor, it is necessary to increase the number of sedimentation tanks or other primary filtration facilities.

5.6.7 Trash rack and screen

Sprinkler irrigation projects often set up a trash rack or trash screen in the following situations:

① When the irrigation water source (such as rivers, reservoirs, or ponds) contains a large quantity of debris (such as litter grottoes or weeds), the trash rack should be established in the upper stream of the pumping station to prevent the debris from going into the sedimentation tank or pool. The trash rack is simple in structure; users can design and manufacture it autonomously corresponding to their actual needs.

② When micro-irrigation water contains a high concentration of suspended particles, filter clogging and cleaning cycles are frequent. In this case, a dirty screen needs to be constructed at the inlet of a pump station or pump immediately. The quantity of mesh in a dirt screen is around 80 to 120. The range is a bit large because the mesh number varies depending on the shape and size of a specific cage. A dirt screen can greatly reduce the concentration of suspended solids, reduce the load of the downstream filter, and improve the efficiency of the filter.

The screen must be cleaned continuously. Otherwise, the screen will be blocked easily, resulting in poor water inflow and affecting the normal operation of the pump.

5.7 Fertilizer apparatus

A fertilizer apparatus is an important component of a micro-irrigation system, which is used to inject the fertilizer solution into the pipe network, and then into the soil in the root area of the crop with water[22-26]. The common fertilizer applications of micro-irrigation systems include a pressure differential fertilizer container, venturi injector, and proportional injector, and hydraulic driven pump.

5.7.1 Pressure differential fertilizer container

The pressure differential fertilizer container is made of corrosion-resistant materials, like plastic or metal [27,28]. It is mainly used for soluble chemical fertilizer. It must be capable of bearing high pressure and have a good sealing performance. At present, the volume of different containers is 10 L, 16 L, 30 L, 50 L, I00 L, and 150 L.

1. Basic structure

The pressure differential fertilizer container is also known as the bypass fertilizer container, which mainly consists of a container, water supply hose, and infusion hose. The water supply hose and infusion hose are connected to the front and back of the regulating valve of the micro-irrigation system; the structure of a pressure differential fertilizer container is shown in Figure 5-29. It is simple, easily manufactured, and does not require any additional power equipment. However, the fertilizer concentration in a pipe network before and after fertilization is largely fluctuant and uncontrollable. So, more fertilizer must be added frequently due to the limited volume of the containers. Also,

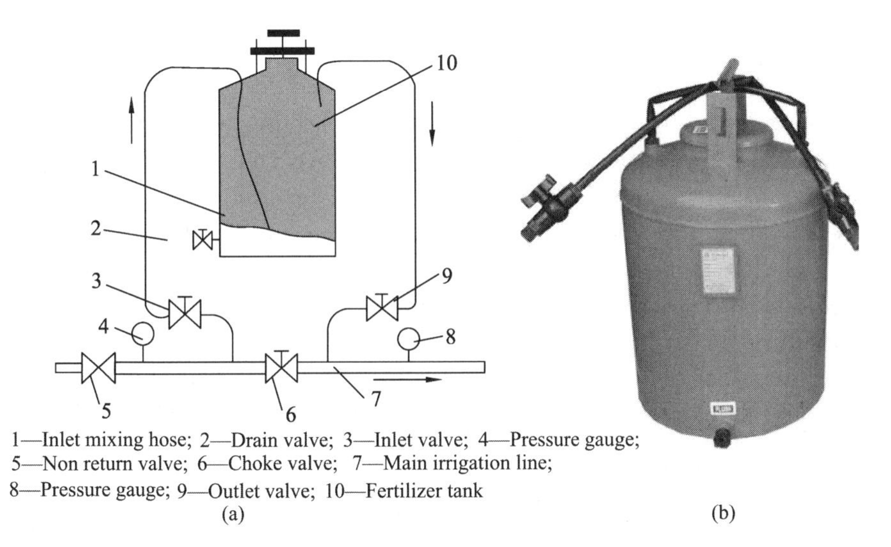

1—Inlet mixing hose; 2—Drain valve; 3—Inlet valve; 4—Pressure gauge;
5—Non return valve; 6—Choke valve; 7—Main irrigation line;
8—Pressure gauge; 9—Outlet valve; 10—Fertilizer tank

(a) (b)

Figure 5-29 Structure diagram of the pressure differential fertilizer container

water head loss must exist due to a lack of a regulating valve. To ensure the compressive strength of the container, it is often made of metal, and a rubber bag is installed in it to prevent the inner walls from corroding, which is called the modified fertilizer container.

2. Working principle

Irrigation water from the water pipe enters the storage container through the water supply hose, mixes with the chemical fertilizer in the container, and then enters the micro-irrigation water supply pipe control valve through the liquid supply hose. After this process, it is sent to the designated field to irrigate through the micro-irrigation pipe. When the system is operating, the flow rate of the fertilizer solution is regulated by adjusting the opening of the valve to control the pressure difference before and after it reaches the valve. When the fertilizer solution is loaded into the rubber bag, the pressure difference is formed by adjusting the valve. This leads to improvement in the efficiency of the fertilizer container. The pressured water from the micro-irrigation pipeline enters the fertilizer container, and the outer surface of the rubber bag is extruded so that the fertilizer enters the water delivery pipe.

At present, there is no national or industrial standards for pressure differential fertilizer containers, so manufacturers can produce them in a way that is most beneficial to their own enterprise's standards. When purchasing, customers should pay attention to whether the technical indicators meet the requirements most suitable for their irrigation process.

5.7.2 Venturi injector

1. Working principle

A venturi injector is a tapered constriction which operates on the principle that a pressure drop accompanies the change in velocity of the water as it passes through the constriction. The pressure drop through a venturi must be sutficient to create a negative pressure. When the flow rate increases, a negative pressure is generated, and the liquid fertilizer is sucked through the suction pipe. From there, it is carried into the pipeline system within the flow of the water[29-35].

The advantages of Venturi injector include: it is a simple structure, there are no actively moving parts, fertilizer solution is absorbed by the open container. The product specifications and models are diverse and the cost is low. One of the disadvantages is the large head loss in the process of suction. For most types, the loss is at least 1/3 of the inlet pressure. Also, the fertilizer device is sensitive to any fluctuation in pressure and flow rate, which will significantly affect the fertilizer mixing ratio (the fertilization uniformity) and the capability to absorb the fertilizer. Also, the operation range of each type of fertilizer application is very narrow, especially when the pressure drop caused by suction is too small or the inlet pressure is too low; the water will flow into the fertilizer container from the main pipe, resulting in the overflow of the fertilizer solution.

2. Improved Venturi injector

A one-way valve can be installed on the Venturi fertilizer application to limit head loss, reduce poor stability, lessen vacuum damage, and fix other possible shortcomings.

It is shown in Figure 5-30. A one-way valve is installed at the fertilizer suction port to prevent the irrigation water in the pipe network flowing into the fertilizer container when the suction of the Venturi fertilizer application is insufficient. A vacuum destruction valve is set up to destroy the local vacuum in the pipeline, preventing the fertilizer solution from being sucked out of the high fertilizer container and into the pipeline when the pipeline empties.

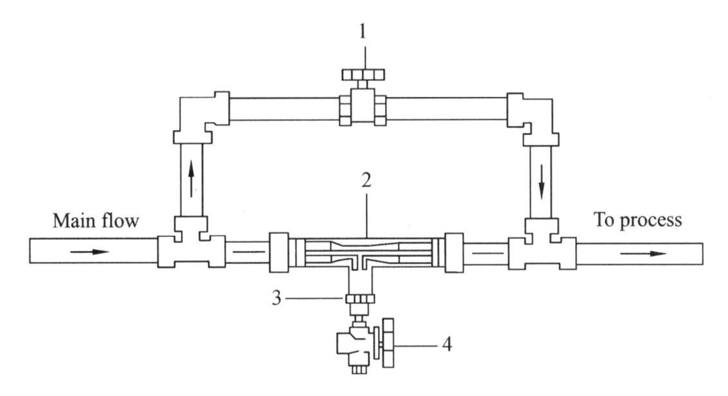

1—Pressure regulating, or flow control valve; 2—Injector; 3—Check valve; 4—Metering valve

Figure 5-30 Venturi fertilizer injector with a one-way valve

5.7.3 Proportional injector

Figure 5-31 shows the schematic diagram of a proportional injector, which is installed directly in the water supply line, operating without an electrical motor[36, 37]. It uses the flow of water as a power source to activate the motor piston, which takes up the required percentage of concentration and injects it into the water. Inside the injector, the injected solution is mixed with water and the water pressure pushes the mixed solution downstream.

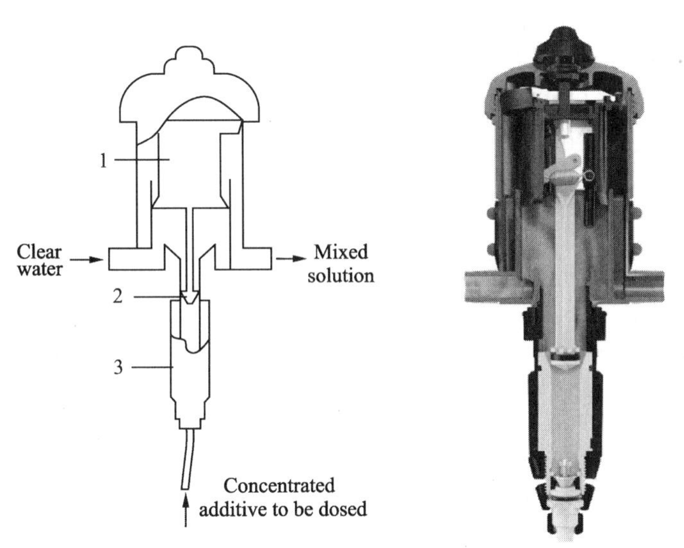

1—Motor piston; 2—Dosing piston; 3—Adjustment injection ratio

Figure 5-31 Diagram of the proportional injector

References

[1] Rural Water Conservancy Bureau of the Ministry of Water Resources, China Development and Training Center for Irrigation and Drainage Technology. Sprinkler and micro-irrigation equipment [M]. Beijing: China Water Conservancy and Hydropower, 1998.

[2] Batchelor C, Lovell C, Murata M. Simple microirrigation techniques for improving irrigation efficiency on vegetable gardens[J]. Agricultural Water Management, 1996, 32(1): 37 - 48.

[3] Barragan J, Bralts V, Wu I P. Assessment of emission uniformity for micro-irrigation design[J]. Biosystems Engineering, 2006, 93(1): 89 - 97.

[4] Cararo D C, Botrel T A, Hills D J, et al. Analysis of clogging in drip emitters during wastewater irrigation [J]. Applied Engineering in Agriculture, 2006, 22(2): 251 - 257.

[5] Wei Q, Shi Y, Dong W, et al. Study on hydraulic performance of drip emitters by computational fluid dynamics [J]. Agricultural Water Management, 2006, 84(1): 130 - 136.

[6] Li G Y, Wang J D, Alam M, et al. Influence of geometrical parameters of labyrinth flow path of drip emitters on hydraulic and anti-clogging performance[J]. Transactions of the ASABE, 2006, 49(3): 637 - 643.

[7] Borssoi A L, Vilas Boas M A, Reisdörfer M, et al. Water application uniformity and fertigation in a dripping irrigation set [J]. Engenharia Agrícola, 2012, 32(4): 718 - 726.

[8] Ali A A M. Anti-clogging drip irrigation emitter design innovation [J]. European international journal of science and technology, 2013, 2(8): 154 - 164.

[9] Lamm F R, Ayars J E, Nakayama F S. Microirrigation for crop production: design, operation, and management[M]. Amsterdam: Elsevier, 2007.

[10] Changade N M, Chavan M L, Jadhav S B, et al. Determination of emission uniformity of emitter in gravity fed drip irrigation system[J]. International Journal of Agricultural Engineering, 2009, 2(1): 88 - 91.

[11] Yuan Z, Waller P M, Choi C Y. Effects of organic acids on salt precipitation in drip emitters and soil[J]. Transactions of the ASAE, 1998, 41(6): 1689 - 1696.

[12] Demir V, Yurdem H, Degirmencioglu A. Development of prediction models for friction losses in drip irrigation laterals equipped with integrated in-line and on-line emitters using dimensional analysis[J]. Biosystems Engineering, 2007, 96(4): 617 - 631.

[13] Ella V B, Reyes M R, Yoder R. Effect of hydraulic head and slope on water distribution uniformity of a low-cost drip irrigation system [J]. Applied Engineering in Agriculture, 2009, 25(3): 349 - 356.

[14] Keller J, Karmeli D. Trickle irrigation design parameters[J]. Transactions of the Asae, 1974, 17(4): 678 - 684.

[15] Li J S, Meng Y B, Li B. Field evaluation of fertigation uniformity as affected

by injector type and manufacturing variability of emitters[J]. Irrigation Science, 2007,25(2): 117 - 125.

[16] Li J S, Du Z, Li Y. Field evaluation of fertigation uniformity for subsurface drip irrigation systems [J]. Transactions of the Chinese Society of Agricultural Engineering, 2008,24(4): 83 - 87.

[17] Kang Y, Nishiyama S.Design of microirrigation submain units[J]. Journal of Irrigation and Drainage Engineering, 1996,122(2): 83 - 89.

[18] Li J, Li Y, Wang J, et al. Microirrigation in China: History, current situation and prospects[J]. Journal of Hydraulic Engineering, 2016,47(3): 372 - 381.

[19] Merriam J L, Keller J. Farm irrigation system evaluation: A guide to management[M]. Logan,Vtah: Utah State University, 1978.

[20] Sun M Y. The research of application situation and technical parameters of drip irrigation[D]. Xianyang: Northwest A&F University,2014.

[21] Solomon K.Manufacturing variation of trickle emitters[J]. Transactions of the ASAE, 1979,22(5):1034 - 1038.

[22] Camp C R. Subsurface drip irrigation: A review[J]. Transactions of the ASAE, 1998,41(5): 1353 - 1367.

[23] Kakhandaki S H, Padmakumari O, Madhusudhan M S, et al. Effect of drip and micro sprinkler irrigation on soil moisture distribution pattern in tomato crop under clay loam soil [J]. International Journal of Agricultural Engineering, 2012,5(1): 121 - 122.

[24] Hoffman G J, Evans R G, Jensen M E, et al. Design and operation of farm irrigation systems[M]. St. Joseph, MI: American Society of Agricultural and Biological Engineers, 2007.

[25] Bracy R P, Parish R L, Rosendale R M. Fertigation uniformity affected by injector type[J]. Hort Technology, 2003, 13(1): 103—105.

[26] Li J, Meng Y, Liu Y. Hydraulic performance of differential pressure tanks for fertigation[J]. Transactions of the ASABE, 2006,49(6): 1815 - 1822.

[27] Meng Y B. Hydraulic performance of injection devices for microirrigation system[M]. Beijing: College of Water Conservancy and Civil Engineering, China Agricultural University.

[28] Neto I E L, Porto R D M. Performance of low-cost ejectors[J]. Journal of Irrigation & Drainage Engineering, 2004,130(2): 122 - 128.

[29] Yuan Z, Choi C Y, Waller P M, et al. Effects of liquid temperature and viscosity on venturi injectors[J]. Transactions of the ASAE, 2000,43(6): 1441 - 1447.

[30] Li B J, Wang X N. Design and experiment on hydrodynamic fertilizer injection unit[J].Journal of Jiangsu University: Natural Science Edition, 2002,23(2): 9 - 12

[31] Kumar M, Rajput T B S, Patel N. Effect of system pressure and solute concentration on fertilizer injection rate of a venturi for fertigation[J]. Journal of Agricultural Engineering, 2012,49 (4): 9 - 13.

[32] Kranz W L, Eisenhauer D E. Calibration accuracy of chemical injection devices[J]. Applied Engineering in Agriculture, 1996,12(2): 189 - 196.

[33] Lancaster M E, Davis J M, Sanders D C. A continuously diluting injector for applying fertilizer to experimental and demonstration plots [J]. Horttechnology, 1998,8(2): 221-224.
[34] Manzano J, Palau C V, Azevedo B M D, et al. Design and installation alternatives of venturi injectors in drip irrigation [J]. Revista Ciencia Agronomica, 2015,46(2): 54-60.
[35] Wang J, Gong S, Xu M, et al. Research and development of liquamatic piston fertilizer pump for micro-irrigation[J]. Transactions of the Chinese Society of Agricultural Engineering, 2006,22(6): 100-103.
[36] Han Q B, Wu W Y, Liu H L, et al. Experiment on fertilizer suction performance of three hydraulic driven pumps[J]. Transactions of the Chinese Society of Agricultural Engineering, 2010,26(2): 43-47.
[37] Yang D S, Li H, Luo Z W. Working principle and performance test of the piston proportion fertilizer applicator[J]. Water Saving Irrigation, 2015,11: 47-50.

CHAPTER 6 Automatic Control Equipment of Sprinkler Irrigation and Micro-irrigation Systems

With the development of economy, water resources, and energy shortage, labor costs rise, more water-saving irrigation systems are being used with automatic control. It has the following advantages [1]:

① It provides timely and appropriate control of irrigation, irrigation scheduling, thereby increasing crop yields and significantly increasing water use.

② It saves labor and cuts operational costs.

③ It is convenient and flexible to arrange work plans, so managers do not have to go to the fields at night or inconvenient times.

④ Due to the increase in effective working hours, it can correspondingly reduce the initial capital investment in pipelines, pumping stations and so on.

6.1 Classification of automation irrigation systems[2]

6.1.1 Full-automation irrigation system

Full-automation irrigation systems do not require manual direction, as it can automatically start and shut down the pump for a long period and automatically irrigate by pre-programmed controls and parameters that reflect crop water demands irrigation specificities. The only manual work that needs to be done is to adjust the control procedures and maintenance control equipment. In this system, in addition to having an emitter, sprinkler, pipes, fittings, pumps, and motor, it also includes central controllers, automatic valves, sensors (for soil moisture, temperature, pressure, water level, rainfall, etc.) and wires.

6.1.2 Semi-automation irrigation system

Semi-automation irrigation systems do not have sensors installed in fields. Irrigation time and schedules are based on pre-programmed procedures, rather than based on the feedback information of crop and soil moisture, and weather conditions to control. The degree of automation of these systems is very different as this system utilizes both a manual and automatic feature, whereas some pumping stations only use automatic controls or only manual. However, even the features of the semi-automation irrigation system itself can vary. For example, some central controllers have a timer with simple programming functions, whereas some do not even have timers, and instead have some sequential switching valves installed.

6.2 Typical components in automation irrigation systems[3]

6.2.1 Central controller

The central controller is seen as "the brain" of the automation irrigation system. The controller sends an electrical signal to the solenoid valve to start and shut down the irrigation system according to the irrigation procedure (irrigation start time, duration, irrigation cycle, etc.) entered by the irrigation manager.

The controller can be categorized as electromechanical (using alternating currents) and mixed road (using direct currents). The capacity of the controller can be large or small; the smallest controller can control a single solenoid valve, and the maximum can control hundreds.

Controller rated input voltage can be 220 V or 110 V, with an average rated output voltage of 24 V. The actual output voltage is usually between 26 and 28 V.

A controller can control several rounds of irrigation. A wheel irrigation area is defined as a station. Manufacturers provide a variety of station controllers for users to choose. For example, Rain Bird Company produced the ESP-MC controller, where there are 8, 12, 16, 24, 32, 40 and many other stations. A station can control 2 to 3 solenoid valves while an ESP-MC controller manages up to 120 solenoid valves.

Before the automatic controller is operated, a time program must be entered. There are several ways to code the program. Rain Bird Company produced the ESP series controller to enable coding through entering the program through keys on the panel. The programs of Irvine produced by Motorola and LINAK produced by Rain Bird are inputted via a separate program editor. A good controller can usually set several methods of programming to satisfy the system of irrigation requirements within different crops or different irrigation methods (sprinkler irrigation, drip irrigation, etc.).

The controller is equipped with public wiring (also known as "zero lines"), terminals, and control wiring (also known as "FireWire") terminals. Also, some controllers also have a sensor interface.

High-quality controllers have a variety of functions, such as:

① Cycle adds infiltration function. An irrigation period is divided into several sections, where each separate time interval can be set to the length desired. This function is important because, in situations where the soil infiltration rate is low, the total daily water demand needs to divide into several intervals to flow to avoid a runoff. It can also be advantageous in situations where there needs to be frequent irrigation, less irrigation (for nursery time of the crops), and when an area is divided into several sections.

② Rainfall delay function. This function can interfere with the original setting procedure after rainfall in an open-air irrigation system. The rain sensor terminates the controller when it rains and delays the program for a certain period. This is helpful so the crops will not be over irrigated and possibly drowned.

③ Manual irrigation control. The general controller has the function to manually start or shut down the system outside of the controller setting irrigation time.

With the development of this technology, the function of the controller will be more and more flexible to make irrigation management more convenient and to satisfy as many production requirements as possible. However, the more the features that are implemented, the higher the price it is, and the more difficult it is to repair. The controller should be selected according to the specific requirements of a given situation.

6.2.2 Automatic valve

There are many types of automatic valves, which can be categorized as hydraulic valves, solenoid valves and so on based on the way the valve operates. The following information explains the working principle of a solenoid valve since it is the most commonly used.

The solenoid valve is generally a diaphragm valve, as shown in Figure 6-1. The solenoid cavity is separated by a rubber diaphragm made specifically for this type of valve. The diaphragm rises underwater pressure and opens the valve, and then closes the valve when water returns to the diaphragm under hydraulic pressure.

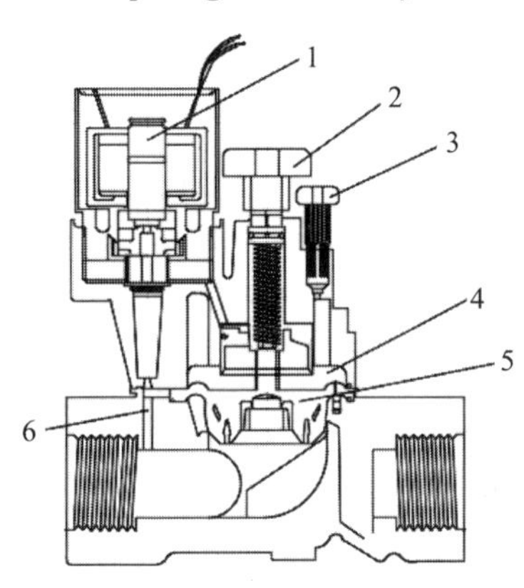

1—Electromagnetic head; 2—Flow rate regulation handle; 3—External exhaust screw; 4—Upper cavity of solenoid valve; 5—Rubber diaphragm; 6—Diversion hole

Figure 6-1 Schematic diagram of a solenoid valve

The upper part of the rubber diaphragm in the solenoid valve has a large area of contact with the water, and the contact area between the lower section and the water is small. If the pressure of the upper and lower diaphragm (water pressure per unit area) is equal. Since the surface area above the diaphragm is larger than the bottom, the water pressure applied to the upper part of the diaphragm is greater than the water pressure below the diaphragm and the diaphragm is pressed back to the diaphragm seat to close the valve. In contrast, if the pressure below the diaphragm is greater than the pressure above the diaphragm, the valve opens. There is a small water hole between the upstream valve and the upper cavity of the diaphragm, so that the upstream water enters the upper cavity and regulates the upstream and downstream water pressure of the diaphragm. The water in the upper cavity of the diaphragm can flow through the upper cavity and the small hole under the electromagnetic head. Thus, there is a small passage between the upstream and downstream sections of the valve, as shown in Figure 6-1. The opening and closing of the channel are controlled by a metal plug on the electromagnetic head. When the metal plug falls, the channel closes and the rise opens. If the channel is opened, the upstream water flows downstream (micro-flow rate), resulting in a smaller diaphragm cavity pressure. On the contrary, if the channel is closed, the water pressure in the upper cavity of the diaphragm will be equal to the upstream pressure of the valve in a short period, so the pressure on the diaphragm will be greater than the pressure under the diaphragm and therefore close the valve.

As shown from the valve opening and closing proccss, its only function is to drive the metal plug in an upwards and downwards movement, which will ultimately block or open the channel between the upstream and downstream sections of the valve. Additionally, it is shown that the driving force of the valve is water pressure. Therefore, when the system flow rate and pressure are insufficient, the solenoid valve is unable to work properly.

The metal plug on the electromagnetic head is lifted by the electromagnetic force and pressed against the spring on the plug.

6.2.3 Wire

After determining the kind of controller and solenoid valve, choosing the according to wire is the next procedure.

The rated voltage of the controller is 24 V. If the wireline is too long (the distance from the most remote solenoid valve to the controller may be more than 1km in some irrigation areas), and the wireline selection is not thick enough, the voltage loss will be too large to reach the minimum voltage required to open the solenoid valve, and the system cannot work properly. So, it is very important to choose the right wires. The wire selection procedure is as follows.

① Determine the allowable voltage loss value. The so-called allowable voltage loss refers to the difference between the output voltage of the controller and the minimum operating voltage of the solenoid valve. If there is no exact data from the manufacturer, it is usually estimated to be 3 V.

② Calculate the allowable maximum resistance value. The calculation formula as follows:

$$R=\frac{U_0}{I} \tag{6-1}$$

where R is the allowable maximum resistance value in Ω, U_0 is the allowable voltage loss value in V, and I is the solenoid valve starting current in A.

③ According to the actual circuit, calculate the allowable resistance value per unit length (such as every 100 m).

④ Compare the findings with the resistance value of the unit length of the wire sold in the market, and determine the suitable type of wire. The principle is that the resistance value of the selected wire should be less than the calculated unit length of allowable resistance.

The distance between the solenoid valve and the controller is 500 m, the output voltage of the controller is 24 V, and the maximum output current is 1 A. The starting current of solenoid valve is 0.45 A, and the minimum operating voltage is 21 V. We decided to adopt the American Geographical lines, various types of unit length (100 m) resistance values as shown in Table 6-1.

$$R=\frac{U_0}{I}=\frac{24-21}{0.45}=6.67\ \Omega \tag{6-2}$$

That is, the maximum allowable resistance on the 500 m wireline is 6.67 Ω, converted to 100 meters per line allowed the value of $\frac{6.67}{500}\times 100=1.33\ \Omega$. No.14 wire should be selected in contrast to Table 6-1.

In practical applications, the valve connection is often complex and cannot simply apply the above method, hence it should be selected according to the manufacturer's instructions.

Table 6-1 Resistance values per unit length of US-made geographical wire

Type	Resistance value/(Ω/100 m)
18	2.23
16	1.40
14	0.88
12	0.55
10	0.35

6.3 Line connection of the automatic control system

Each solenoid valve has two lines, one of which can be used as a control line (also known as "firewire"), and the other can be used as a public line (also known as "zero lines"). Each control line is connected to a terminal of the corresponding station of the controller. There is only one public line in the system, which can be connected to the common terminal (Figure 6-2).

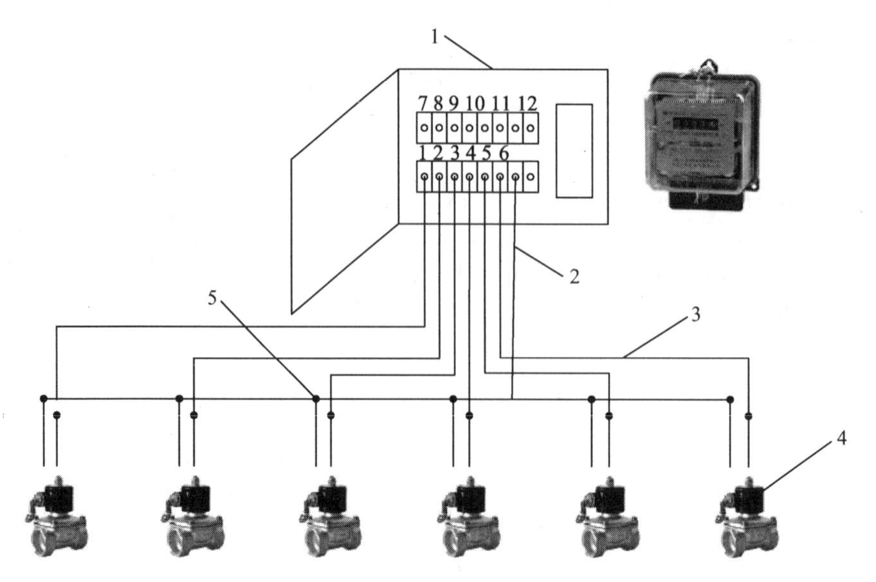

1—The central controller(6 Station); 2—Public lin(Zero line); 3—Fire; 4—Solenoid valve; 5—Waterproof joint

Figure 6-2 Line connection of a typical automatic control system

References

[1] Luo J Y. Theory and technology of water saving irrigation[M]. Wuhan: Wuhan university press, 2003.

[2] Zheng Y Q, Li G Y, Dang P. Sprinkler irrigation and micro irrigation equipment[M]. Beijing: Water conservancy and Hydropower Publishing House, 1998.

[3] Zhao Y D, Zhang J, Wang H L. Precision water saving irrigation control technology[M]. Beijing: Electronic Industry Press, 2010.